Studies in Computational Intelligence 462

Editor-in-Chief

Prof. Janusz Kacprzyk
Systems Research Institute
Polish Academy of Sciences
ul. Newelska 6
01-447 Warsaw
Poland
E-mail: kacprzyk@ibspan.waw.pl

T0122615

For further volumes:
http://www.springer.com/series/7092

Anne Håkansson and Ronald Hartung (Eds.)

Agent and Multi-Agent Systems in Distributed Systems – Digital Economy and E-Commerce

 Springer

Editors
Assoc Prof. Anne Håkansson
KTH Royal Institute of Technology
Software and Computer Systems
Kista
Sweden

Prof. Ronald Hartung
Department of Computer Science
Franklin University
Columbus
USA

ISSN 1860-949X
e-ISSN 1860-9503
ISBN 978-3-642-42797-8
ISBN 978-3-642-35208-9 (eBook)
DOI 10.1007/978-3-642-35208-9
Springer Heidelberg New York Dordrecht London

Printed on acid-free paper

Springer is part of Springer Science+Business Media (www.springer.com)

Editorial

Anne Håkansson and Ronald Hartung

This is a book on agents and multi-agent systems in distributed systems applied to the digital economy and in e-commerce. It represents a flora of technologies and applications of systems within the business systems domain. Over the last two decades, business has become increasing dependent on and involved with electronic systems. Not merely as data processing for the accounting of business activity, but as a means of improving manufacturing and doing business by selling and buying all sorts of products and services, globally. Moreover, information is becoming as much of a product as durable goods.

The business community has transformed from the view of electronic business as a separate discipline to a world where almost no business entity can ignore the electronic side of their enterprise. In early stages of adoption of these systems, the systems remained under human controls. Now, the shift is increasing moving to fully automated system, running with minimal human supervision. The shift will be seen in the extension of "just-in-time" methods into a whole new dimension both in time and space. Just-in-time methods came into being to manage inventory of goods. The next level enabled by electronic business will allow companies to ensure the right products with right price at the right delivery time in a global market.

The transformation of business by electronic systems is progressing and, indeed, it is speeding up. The authors in this book are striving to enable the transformations and to look ahead to where this will lead. The challenges are large, but the possibilities are even larger. The future looks exciting and it will be driver for all kinds of systems using agent and multi-agent technology.

The book combines the fields of agent systems and businesses and is based on the concept of developing and applying agent technology and distributed systems to the business genre. Hence, the book becomes a platform to share ideas and presents research in agent technology in the field of economy and commerce and application to real problems.

Research in multi-agent systems offers promising support for solutions in distributed systems, for digital economy and ecommerce. In digital economy, digital business is carried out via digital technologies or channels and is a global network of economic and social activities. Commonly, digital economy refers to business trading and services, but it also encompasses all aspects of life ranging from health, education, and business to banking. The digital domain has become an electronic mirror of the physical world, inhabited by electronic second selves. In ecommerce, or electronic commerce, products and services, or information, are bought, sold, or exchanged using digital networks via technologies such as electronic transfers, marketing, transactions, and interchanges. The ecommerce covers Business-to-Business, Business-to-Consumer, and Consumer-to-Consumer, but other possible forms as well, such as, Business-to-Government and social commerce.

The chapters range over both technology and applications, illustrating the possible uses of agents in an enterprise domain, and design and analytic methods, needed to provide the solid foundation required for practical systems. The organization of the chapters is arranged into several areas, however the ideas in the chapters overlap between the analytic work and the applications themselves.

The first three chapters of this book present solutions for the digital economy giving samples of efforts improving the production and distribution by using agents and multi-agent technology in distributed environments. These chapters discuss improvements in production, inventory and transportation planning, e-sourcing clusters in network economy and knowledge exchange between agents applicable to online trading agents.

Next chapters give security solutions to both digital economy and e-commerce. The first chapter focuses on security for systems that work in economy and e-commerce environment and the second chapter focuses more on e-commerce. More specifically, the security concerns defending systems from network attacks and preserve network traffics by preventing malicious codes from sneaking into systems and avoiding implementing security problems in the computer systems.

From these first four chapters, the focus moves towards e-commerce. The following chapters of the book present mapping and alignment of ontologies for business, negotiation, automated auctions, recommender systems to support traders in business activities, and simulations for games.

The contributors to the book were carefully selected authors with peer-reviewed chapters. The authors are academics, educators, and practitioners who are doing research in agent and multi-agent systems in distributed systems for digital economy and e-commerce to meet, improve, and overcome challenges of the digital economy and e-commerce sphere. The agent and multi-agent solutions are applied in implementing real-life, exciting developments associated with the need to eliminate problems of distributed systems.

The opening chapter of the book, Chapter 1, *Improving multi-actor production, inventory and transportation planning through agent-based optimization* is authored by Johan Holmgren, Jan A. Persson and Paul Davidsson.

In this chapter, the authors present an agent-based optimization approach to improve multi-actor production, inventory and transportation planning. The agent-based optimization approach is built upon the principles of Dantzig-Wolfe column generation to optimize production, inventory, and transportation and improve planning for actors involved in supply chain. The agent-based optimization approach can also be used as strategic decision support to show how the involved actors may benefit from applying Vendor Managed Inventory (VMI).

The authors use an agent-based distributed system solution where agents run on different computers, to increase confidentiality, robustness and provides the possibility of parallel computing, at the expense of increased processing and communication. The parallelization significantly improves time performance and allows solving more complex sub-problems, making it possible to capture more details of the actual problem and obtain solutions with higher quality.

The agent-based optimization approach is applied to a real-world integrated production, inventory and routing problem, and the results indicate that an increased number of VMI customers may significant reduce the total cost in the system.

The next chapter, Chapter 2, titled *e-Sourcing Clusters in Network Economy*, is authored by Konrad Fuks, and Arkadiusz Kawa

In this book chapter, the authors present a new resources acquisition method and an e-sourcing cluster concept. The authors use digital business and agent technology to merge clusters, which support companies to more easily and speedier access more attractive sources, respond to the increased globalization and cost reduction, as well as, transform and boost individual competitiveness and take advantage of the synergy effects.

The proposed model of e-Sourcing clusters in network economy, eSoC, uses agents to significantly improve the company's resource acquisition process which results in a lower risk factor of investing into a new business model. This can lead to better cooperation, cross-companies relations and collaboration in achieving goals. Four different strategies are presented in the chapter. The selection of strategies depends on the level of agents' empathy and the distribution methods and the choice of the best strategy is influenced by the size of resource and of the demand of the enterprises.

A simulation model of four e-sourcing cluster strategies is built and verified. The results show tendencies that some strategies can be more beneficial than others depending on cluster demands and supply.

The chapter that follows, Chapter 3, is titled *Formalization and verification of knowledge management in digital economy* and is authored by Lilia Georgieva and Imran Zia.

The authors propose a framework for modelling knowledge exchange between agents, applicable to online trading agents. The aim is to model a knowledge exchange check among agents that, consequently, checks correctness of properties of knowledge management processes. This includes identifying fundamental knowledge streams essential for knowledge sharing, exchanging among agents in digital economy and proving that the exchange is correct.

The framework relies on identification of processes supporting knowledge management, which are determining knowledge and available knowledge, determining knowledge gaps, knowledge locks, knowledge sharing, knowledge utilisation and evaluation.

In the chapter, the authors express knowledge exchange in epistemic logic. The formalisation is encoded in an agent modelling protocol language, which becomes verifiable by model checking.

The next chapter, Chapter 4, is titled *Agent-based Artificial Immune Systems (ABAIS) for Intrusion Detections: Inspiration from Danger Theory* and authored by Chung-Ming Ou, C.R. Ou and Yao-Tien Wang.

The chapter presents an agent-based system for intrusion detections used for both digital economy and e-commerce. The authors facilitate intelligent agent mechanisms with artificial immune systems based on danger theory to improve intrusion detection systems. The dendritic cells agent can effectively reduce both the false positive rate and the false negative rate of danger signals issued by computer hosts. Three agents are coordinated to exchange information of intrusion detections.

The solution is based on an Agent-based Artificial Immune System, ABAIS, where the agents are coordinated to exchange information of intrusion detections. The intelligence behind ABIDS is based on the functionality of dendritic cells in human immune systems. Antigens are profiles of system calls while corresponding behaviours are regarded as signals. ABAIS is based on the danger theory while dendritic cells agents are emulated for innate immune subsystem and T-cell agents are for adaptive immune subsystem. This ABIDS is based on the dual detections of dendritic cells agent for signals and T-cell agent for antigen, where each agent coordinates with other to calculate danger value. According to danger values, an immune response for malicious behaviours is activated by either computer host or Security Operating Center (SOC).

The following chapter, Chapter 5, titled *Enhancing security of e-commerce software using multi-agent systems* is authored by Esmiralda Moradian.

In this chapter, the author presents a solution to enhance security of e-commerce software using multi-agent system. The ambition is to prevent a number of cybercrimes and simple mistakes, such as, not insuring that all traffic in and out of a network passes through firewall, by considering security of e-commerce systems from the very beginning, i.e., early stage of the e-commerce software development.

The research, in this chapter, focuses on providing support to engineers during system development process to avoid security problems. A multi-agent system is applied to support implementation of patterns and extracting security information, and provide traceability in the engineering process.

The following chapter, Chapter 6, titled *Conceptual Ontology Intersection for Mapping and Alignment of Ontologies* is authored by Anne Håkansson and Dan Wu.

In this chapter the authors present a conceptual ontology intersection to map and align contents of the ontologies. This intersection is a conceptual ontology bridge between involved ontologies and contains parts from these ontologies. The contents of the ontologies are extracted by using agents for syntactic mapping and synonym alignment with the support by ontology repository, a rule base and a synonym lexicon. The result is a set of concepts that together constitute the intersection, which is used for combining new incoming ontologies and, thereby, providing complex services.

Contents of the ontologies are extracted with the help of agents and compared by syntactic matching and syntactic mapping, as well as, synonym alignment. The syntactic mapping and synonym alignment are applied on ontologies fetched from the Web. The concepts and relations are matched and mapped by using a rule base, and an ontology repository. During the mapping process, a meta-model of the ontologies is built. The meta-model is a conceptual ontology intersection, which is an ontology bridge containing syntactic similarities between the ontologies. To enrich the connections between the ontologies, synonym alignment is used to find synonyms in the ontologies. The alignment uses a synonym lexicon and the result expands the meta-model with synonyms, i.e., the conceptual ontology intersection. By this the ontology intersection provides contexts for the ontologies. Building contexts for the ontologies can help to companies to use them for digital economy and e-commerce.

The chapter that follows, Chapter 7, is titled *Implications and solution for high-speed business architecture* and authored by Ronald L. Hartung.

This chapter presents a forward-looking view of extremely rapid business operations, which is based on the trajectory of business operations. The work is based on the continued acceleration and the distribution of manufacturing and production into a worldwide network. Also, it is based on the opportunity to use custom manufacturing.

As a solution, the chapter presents multi-agent negotiation for a set of interrelated commodities as a possible valuable business process. The multi-agent negotiation makes the purchase and delivery of supplies, the scheduling of production and the coordination of customer orders and delivery a single unified process. The research shows an interesting view for systems that the IT professionals will build and manage. An exciting aspect of this research is the reversal of the standardization of the industrial revolution, where in the future can become a world of custom items tailored for the individual. Moreover, high-speed delivery and automated operation can offer a lower cost point and business efficiency.

The next chapter, Chapter 8, is titled *Mathematical Models of Automated Auctions* and authored by Erol Gelenbe and Kumaara Velan.

The authors of this chapter present models of automated e-commerce activities under the formal rule of automated auctions. The models are developed to study auctions as economic mechanisms that have well-defined protocols and rules to ensure timely conclusions and principled outcomes.

The contribution of this chapter is illustrating how probabilistic modelling techniques that are popular among the operations research community can be used to study automated auctions. Tools and techniques from stochastic modelling theory are employed and the authors show how these techniques can be used to build auction models of increasing complexity using simple first principles. In the models, the auction proceeds in an ascending price order and the process is represented as a jump process that takes valuation from a discrete state-space. Performance measures of a designated Special Bidder are obtained with respect to the rest of the auction participants. The bidder's optimal speed at which bids are placed in the auction, so called bid rate, with respect to the bid rate of the other bidders and the urgency of the seller to make a sale is derived.

The following chapter, Chapter 9, is titled *Generating B2C Recommendations Using a Fully Decentralized Architecture* and authored by Domenico Rosaci, and Giuseppe M.L. Sarné.

The chapter presents a recommender system, called Trader REcommender System (TRES). The system generates content-based and collaborative filtering suggestions as support traders in their B2C activities. The suggestions take the customers' interests and preferences in B2C activities into account and contribute to increase the performances of the E-Commerce sites. The TRES system is based on a multi-agent architecture where agents are used to make the merchants capable of generating effective and efficient recommendations. The architecture supports generating recommendations without requiring help of any centralized computational unit, which makes the system scalable, with respect to the size of the users' community, and preserve privacy of the customers.

Experimental simulations using a JADE-based prototypal implementation over an agent platform gives good results. The results show that the TRES system's effectiveness in generating suggestions is higher than by the TRES algorithm, implemented without considering the different trading phases enacted by a B2C process, and other competitors' recommenders.

The last chapter, Chapter 10, is titled *Simulation analysis using multi-agent systems for generalized matching pennies games* and authored by Ichiro Nishizaki, Tsuyoshi Nakakura and Tomohiro Hayashida.

In this chapter, the authors investigate the long-run behaviour of players in the generalized matching pennies game by employing an approach based on adaptive behavioural models. An agent-based simulation system is developed in which artificial adaptive agents have mechanisms of decision making and learning based on neural networks and genetic algorithms.

The authors analyse strategy choices of agents and the obtained payoffs in the simulations, and compare the predictions of Nash equilibria, the experimental data, and the results of the simulations with artificial adaptive agents. Also, the authors examine similarities between the behaviours of the human subjects in the experiments and those of the artificial adaptive agents in the simulations.

Anne Håkansson
KTH Royal Institute of Technology, Kista, Sweden

Ronald Hartung
Franklin University, Columbus, Ohio, USA

Contents

Improving Multi-actor Production, Inventory and Transportation Planning through Agent-Based Optimization

Johan Holmgren[1], Jan A. Persson[1], and Paul Davidsson[1,2]

[1] School of Computing, Blekinge Institute of Technology, SE-374 24 Karlshamn, Sweden
[2] School of Technology, Malmö University, SE-205 06 Malmö, Sweden
{johan.holmgren,jan.persson,paul.davidsson}@bth.se

Abstract. We present an agent-based optimization approach that is built upon the principles of Dantzig-Wolfe column generation, which is a classic reformulation technique. We show how the approach can be used to optimize production, inventory, and transportation, which may result in improved planning for the involved supply chain actors. An important advantage is the possibility to keep information locally when possible, while still enabling global optimization of supply chain activities. In particular, the approach can be used as strategic decision support to show how the involved actors may benefit from applying Vendor Managed Inventory (VMI). In a case study, the approach has been applied to a real-world integrated production, inventory and routing problem, and the results from our experiments indicate that an increased number of VMI customers may give a significant reduction of the total cost in the system. Moreover, we analyze the communication overhead that is caused by using an agent-based, rather than a traditional (non agent-based) approach to decomposition, and some advantages and disadvantages are discussed.

1 Introduction

Supply chain management is an area in which actors may experience great potentials by the use of efficient e-business solutions [1]. The introduction of powerful computers and efficient methods for formulating and solving complex optimization problems has made it possible to improve the operations in supply chains. In traditional central approaches for solving optimization problems, all information that need to be used when formulating and solving a problem has to be shared with a central node of computation. However, if multiple organizations are involved there is often a wish to keep sensitive information local. We propose an agent-based approach for integrated optimization of production, inventory and transportation, which has the potential to offer increased confidentiality for the involved organizations.

In this paper we describe how Dantzig-Wolfe (DW) decomposition [2], which is a classic reformulation technique, can be incorporated in a multi-agent system. The main purpose is to give a detailed account to how a classic optimization approach can be "agentified". Another purpose is to validate that some positive characteristics can be achieved by using this type of approach. The main characteristic is confidentiality of information, which is of particular importance in applications where different, potentially

A. Håkansson & R. Hartung (Eds.): *Agent & Multi-Agent Syst. in Distrib. Syst.*, SCI 462, pp. 1–31.
DOI: 10.1007/978-3-642-35208-9_1   © Springer-Verlag Berlin Heidelberg 2013

competing, organizations are represented. Moreover, we investigate the possibility to achieve performance improvements by distribution and parallelization of the approach.

In a case study we apply our approach to a real-world integrated production, inventory and routing problem. The problem includes production planning, vehicle routing and inventory planning, and the objective is to minimize the total cost for production, transportation and inventory holding, while meeting the customers' demands for products. In particular, we show how the approach can be used for strategic decision making by quantifying the economic benefits that can be achieved by introducing *Vendor Managed Inventory* (VMI) for one or more supplier-customer relations. In VMI [3], the supplier is responsible for replenishing the customers' inventories, while given continuous access to information about forecasted customer demand and storage levels. Typically, the supplier owns the products until the customer removes them from inventory. The supplier benefits from being able to get updated information about forecasted customer demand and storage levels in contrast to being forced to deal with often late arriving, and changing customer orders. This provides flexibility in production and transportation planning. The customer potentially benefits from being able to pay later (and typically less) for products. Also, since it is the responsibility of the supplier to decide about deliveries, the customer does not have to bother about ordering products. For the studied problem we have performed simulation experiments, which indicate that an increased number of VMI customers may give a significant reduction of the total cost in the system. Moreover, we discuss how the approach can be used for operational planning, e.g., by providing decision support for real-world planners concerning how to perform supply chain activities.

In summary, we focus on the supply chain cooperation dimension of e-business by: (1) exploring an approach for integrated planning (involving multiple actors), which supports the keeping of planning related information confidential, and (2) from a strategic perspective, analyzing and quantifying the benefit of increased cooperation by VMI.

In next section we introduce the reader to the fields of agent-based optimization and decomposition, and in Section 3 we given an account to integrated planning of production, inventory and transportation, including some related work. In Section 4, we present a real-world case problem, and for the case problem, in Section 5 we describe an agent-based decomposition approach. Finally, in Section 6 we present some computational experiments before concluding the paper with a discussion on confidentiality in Section 7 and some conclusions and directions for future work Section 8.

2   Agent-Based Optimization and Decomposition

Agent-based approaches to optimization can be built in many ways, e.g., by using classical agent concepts, such as auctions and negotiation, as in the examples provided by Karageorgos et al. [4] and by Dorer and Calisti [5]. However, the focus here is on agent-based approaches that make use of techniques and concepts from classical optimization, e.g., methods for formulating and solving complex optimization problems.

It has been argued that the strengths and weaknesses of agent-based approaches and classical optimization techniques complement each other well for dynamic resource allocation problems [6]. The strengths and weaknesses of agent-based approaches and mathematical optimization techniques were compared for resource allocation in the domain of production and transportation. The comparison indicated that agent-based approaches tend to be preferable when: the size of the problem is large, communication and computational stability is low, the time scale of the domain is short, the domain is modular in nature, the structure of the domain changes frequently and there is sensitive information that should be kept locally, and classical optimization techniques when: the cost of communication is high, the domain is monolithic in nature, the quality of the solution is important, and it is desired that the quality of the solution can be guaranteed. Moreover, the comparison indicated that the properties of the two approaches are complementary and that it can be advantageous to combine them. In a case study, two hybrid approaches were tested:

1. Optimization was embedded in the agents to improve their abilities to take good decisions.
2. Optimization was used for creating long-term coarse plans, which were refined dynamically by the agents.

Another hybrid approach is referred to as *distributed constraint optimization*. According to Petcu [7], a *Constraint Optimization Problem* (COP) is defined as a set of variables with corresponding discrete and finite variable domains and a set of utility functions. Each utility function assigns a utility to each possible assignment of the variables, and the purpose is to find the variable instantiation that maximizes the sum of utilities of the utility functions. An interesting property is that variable domains are not restricted to numerical values, e.g., a variable domain may refer to colors or whatever is relevant for a particular problem. A *Distributed Constraint Optimization Problem* (DCOP) is defined as a set of agents, where each agent owns a centralized COP (i.e., a local subproblem), and a set of inter-agent utility functions, which are defined over variables from the local subproblems. An inter-agent utility function represents the award that is assigned to the involved agents when they take a joint decision.

Decomposition approaches, such as Dantzig-Wolfe column generation [2] has been developed for solving linear problems, and Lagrangean relaxation [8] and Benders' decomposition [9] have been developed for solving *Mixed Integer Linear optimization Problems* (MILPs). Moreover, decomposition approaches for linear problems have been used together with *branch-and-price* [10] to solve MILPs.

Combining agents with decomposition approaches is another, relatively new approach, which we find particularly interesting to investigate. An example is provided by Hirayama [11], who proposed an agent-based approach for solving the *Generalized Mutual Assignment Problem*. In the approach, which was built using a distributed solution protocol based on Lagrangean decomposition and distributed constraint satisfaction; agents were used to solve individual optimization problems, which were coordinated in order to improve a global solution.

In this paper the focus is on another decomposition approach, and we describe how Dantzig-Wolfe decomposition can be agentified into a multi agent system. In Dantzig-Wolfe decomposition, a linear *Master Problem* (MP) is reformulated into a *Restricted Master Problem* (RMP) containing only a subset of the variables in MP, and a set of subproblems which produce new solutions (columns) that are coordinated by RMP. In an iterative process, subproblem solutions based on so-called dual variables (which is a control mechanism) are added as improving variables to RMP. A dual variable is often interpreted as a value or price, e.g., for obtaining an additional unit of a scarce resource, or for producing one more unit of a particular product type.

In contrast to approaches to agent-based optimization, in which agents are able to communicate directly with each other in a peer-to-peer fashion, agent-based decomposition requires a coordinator agent who is responsible for managing the problem solving process. In DW decomposition, the coordinator agent corresponds to the RMP, and planner agents are responsible for providing plans (solutions) to the coordinator by solving subproblems. A conceptual illustration of agent-based DW decomposition is provided in Fig. 1, in which it can be seen that the coordinator sends dual variables to the planner agents, who return plans to the coordinator.

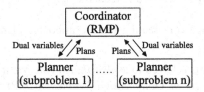

Fig. 1. A conceptual model of agent-based Dantzig-Wolfe decomposition. In an iterative process, the coordinator sends dual variables to the planner agents, who return improving plans to the coordinator.

For a case problem, in Section 5 we provide a detailed example of an agent-based decomposition approach for an integrated production, inventory, and routing problem. We have chosen a decomposition formulation that we find attractive, in particular since it allows for a natural interpretation of dual prices and subproblems. The studied problem class captures the difficulties with distributed decision-making since information and resources typically are distributed and the exact conditions, e.g., the demand and the availability of resources, are unknown in advance.

Planning tasks in supply chains are often performed by different organizations, which is why confidentiality is an important concept. Traditionally, a decomposition algorithm runs in a single process (on a single computer), which needs access to all information that is required for formulating and solving the optimization problem. In such an approach it is typically not possible to achieve confidentiality due to the fact that information, which might be considered sensitive for the planners may have to be shared. With an agent-based approach, where different problems are represented by different agents, it is often possible to run the optimization with less need for sharing sensitive information. In our case problem there is an agent who coordinates production, transportation and inventory. The coordinator agent needs access to customer demand forecasts, but it does not need to know any underlying details about how transportation plans and production plans are created. Hence, the agents responsible for solving subproblem do not need to share all information to the coordinator. Obviously, the use of agents may have a negative impact on the performance, i.e., concerning the execution time due to an increased need for communication. However, a possible advantage over classic decomposition approaches is that it is straightforward to distribute an agent system over several computers. In a distributed solution approach, there is a potential for reduced computation time and in some cases improved solution quality since more computing power allows for solving more complex subproblems. The reason is that the subproblems can be solved in parallel, which is impossible when the approach runs on a single processor. Further, a decentralized approach makes the system less vulnerable to single point failures. In the agent-based approach, where the coordinator agent typically retains control of all decisions, a failure of the coordinator agent is fatal. However, in case of a planner agent failure, the rest of the planners will still be able to produce new (improving) plans, which can be considered by the coordinator.

3 Integrated Planning of Production, Inventory and Transportation

Raw material supply, production and transportation have often been separated by large inventory buffers allowing different supply chain activities, such as production and transportation, to be planned separately. Various planning problems for different parts of the supply chain have been studied; a survey of lot sizing and scheduling problems is provided by Drexl and Kimms [12], and an overview of the vehicle routing problem and variations is given by Toth and Vigo [13]. The importance of inventory reduction has led to an increased interest in integrated planning of different logistical activities [14], and a review covering efforts in the area of integrated supply chain planning is provided by Sarmiento and Nagi [15].

Our focus is on optimization approaches that consider planning of production, inventory, as well as transportation. There is research focusing on only two of these aspects (e.g., [16,17,18]). According to our knowledge, the earliest contribution that combines all three problem aspects, i.e., planning of production, inventory and transportation, was presented by Chandra and Fisher [19]. For a multi-period planning horizon, the authors solved a combination of the production scheduling problem and the vehicle routing problem where multiple products were distributed from a single production facility to a number of customers.

A more recent approach was presented by Lei et al. [20], who considered an integrated production, inventory, and routing problem with a single product, multiple heterogeneous production plants, multiple customer demand centers and heterogeneous vehicles. Inventory management was considered both at the production plants and at the customer demand centers. The model was approached by formulating a large mixed-integer linear problem, which was solved using a 2-phase solution method. In phase one, the problem was reformulated to only include direct transportation between the producers and the customer demand centers. This restricted problem gives a feasible but non-optimal solution to the original problem. In phase two, a heuristic approach was used to improve the solution from phase one by also considering transport routes involving multiple customer demand centers. The main differences compared to their approach are that we model transportation in more detail, and that we use a different solution approach.

Another recent contribution is provided by Persson and Göete-Lundgren [21], who formulated and solved an optimization model for planning the production at a set of oil refineries and shipments of finished products while considering inventories both at producer and customer depots. Their problem is similar to the one we are considering. However, they used a longer planning horizon, but with longer individual time periods, there was no separation of the production planning into subproblems, and they used ships for transportation whereas we use trucks including explicit modeling of driving time restrictions. They proposed a solution approach based on column generation and valid inequalities, and integer solutions were obtained by using a fixing strategy in which vehicles were fixed to visit certain depots at specific times. To determine how much of each product should be delivered to each depot, an integer model was formulated and solved.

Bilgen and Ozkarahan [22] considered a problem for optimization of blending and shipment of grain (bulk) products from a set of producers to a set of customers. Whereas our approach is built on decomposition, Bilgen and Ozkarahan formulated a single MILP model with the objective to minimize the costs for blending, loading, transportation and inventory. The optimization problem was solved with the ILOG CPLEX solver for a rather short planning horizon.

4   Real World Case Problem Description

We consider a real-world case problem with a producer of vegetable oils and a single hauler that manages a fleet of bulk trucks that take care of deliveries of finished products to a set of customers. Production, and hence the planning of production and transportation is driven by the arrival of customer orders. Production is performed on multiple production lines, which are scheduled (individually) to match the shipping times that are chosen to match the delivery time windows of the customer orders. A typical horizon for production and transportation planning is usually less than a week.

Before loaded onto vehicles, finished products are stored in short-term inventory at the producer depot. The capacity of this short-term inventory is rather limited, and shipping is typically initiated the same day as the last production step is finished. At delivery, products are stored in customer inventories. Starting with a full (or close to full) truck load at the producer, sometimes a truck visits only one customer before returning but sometimes deliveries are grouped together and the truck visits multiple customers in a trip. Occasionally, a vehicle can be scheduled for a non-empty transport on the return trip from a customer to the producer. When the transportation demand exceeds the available transportation capacity, it is possible to call in extra capacity to a higher cost. Furthermore, time and costs for loading and unloading are considered, and the drivers have to follow the European Economic Community (EEC) regulations (EEC 3820/1985) for working and resting hours.

Raw material arrive to the production plant by boat or truck in quantities based on long term forecasts since the order-to-delivery lead time of raw material considerably exceeds the production planning horizon. Therefore, an unlimited supply of raw material can be assumed, and the considered operational production planning can be separated from ordering of raw material.

Currently, the considered real-world producer is considering an introduction of VMI for some of its customers. An introduction of VMI requires that accurate forecasts of consumption can be made available, which is also assumed to be possible. VMI might lead to a higher flexibility in the production and transportation planning, and the expectation is that a higher utilization of (often limited) production and transportation resources can be achieved.

In Fig. 2 we provide a conceptual illustration of the case problem with a small example of a transportation network containing a factory, inventories, customer depots and trucks, and planners who are responsible for taking decisions about physical resources.

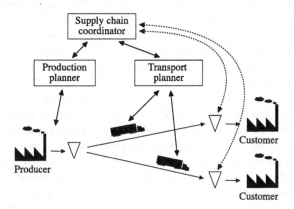

Fig. 2. An illustration of the case problem, with physical resources, planners and a supply chain coordinator. The dashed arrows represent connections that are present only for VMI customers.

5   Decomposition Formulation

The identified real world problem is modeled as a MILP, which is built using the principles of Dantzig-Wolfe decomposition. The decomposition formulation includes a linear DW master problem, a set of transportation subproblems and a set of production scheduling subproblems for construction of transportation plans and production plans. Even though MP is a linear problem, there are no restrictions regarding which types of variables the subproblems are allowed to model. It should be mentioned that DW decomposition is a technique for reformulating optimization problems, and in many cases it is necessary to apply a branch and bound method, i.e., branch-and-price [10], to be able to use it for solving problems with integer variables.

In addition to the practical issues, like for example confidentiality, reasons for not formulating the problem as a large MILP include: 1) a large MILP would most probably be difficult, or impossible, to solve to optimality due to its size and high complexity, and 2) a decomposition approach allows the subproblems to be reformulated without modifying the master problem.

It is possible to use alternative decomposition approaches for modeling the studied problem, e.g., Lagrangean relaxation [8] and Benders' decomposition [9]. In our opinion, the choice of which decomposition technique to use can often be seen as a matter of preference. Important reasons for using DW decomposition for modeling the studied

problem is that it allows us to benefit from the special block structure that characterize the problem. According to Fumero and Vercellis [23], problems in the studied area
tend to use an underlying network structure that may be exploited by decomposition
approaches. The objective is to minimize the costs for production, distribution, and inventory holding, while satisfying the customers' product demand throughout a given
planning horizon. The problem formulation includes production planning at producer
depots, transportation planning including route choice, quantities to ship and times for
deliveries to customers and pickups from the producers, and inventory planning of finished products. The presented approach focuses on the VMI situation due to its potential
for improved resource utilization. However, it is not limited to VMI since a tight specification of customer inventory constraints mimics a non-VMI situation.

We have built the agent-based decomposition approach using a hierarchical agent
model for decision making in a multi-agent-based supply chain simulation model called
TAPAS, which was presented by Davidsson et al. [24]. The model contains a supply
chain coordinator, a transport buyer, a product buyer, production planners, transport
planners, and customers. However, in the suggested decomposition approach we have
chosen not to model the product and transport buyer agents. The downscaled version
of the hierarchical agent model in TAPAS corresponds to the agent model in Fig. 2. In
the agent system, the supply chain coordinator agent represents MP, each production
planner agent handles a set of production scheduling subproblems (one subproblem for
each production line in one producer depot), and each transport planner agent handles a
set of transportation subproblems (one subproblem for each vehicle in its vehicle fleet).

In the problem formulation, we let D^P denote the set of producer depots, D^C the set
of customer depots, $D = D^P \cup D^C$ the set of all depots, V the set of (inhomogeneous)
vehicles, P the set of product types, and L the set of production lines. The planning
horizon is represented by an ordered set $T = \{1, 2, \ldots, \bar{t}\}$ of discrete time periods with
uniform length τ. A *transportation plan* for a vehicle is defined as the amount of each
product delivered to each customer depot and picked-up from each producer depot in
each time period throughout the planning horizon. The set of all feasible transportation
plans for a vehicle $v \in V$ is denoted R_v, and the cost for using plan $r \in R_v$ is denoted
ψ_r. If d is a producer depot (i.e., $d \in D^P$) we let variable x_{dptr} denote the amount of
product p that is picked-up from depot d for plan r by vehicle v in period t, otherwise
$(d \in D^C)$ x_{dptr} represents a delivery to depot d. Similarly, we define a *production plan*
for a production line as the amount of each product that is produced in each time period
throughout the planning horizon. We let S_l denote the set of all valid production plans
for production line $l \in L$, and ω_s denotes the cost for using plan $s \in S_l$. For production
plan $s \in S_l$, which represents a production line located at some depot $d = d(s)$, we let
y_{dpts} represent the amount of product p produced in period t. Furthermore, parameter
ϱ_{dpt} denotes the demand for product $p \in P$ at customer depot $d \in D^C$ in time period
$t \in T$. Hence, the parameter ϱ_{dpt} specifies the amount of product p removed from
the customer inventory in time period t. It should be noted that, in our methodological
approach, only a subset of all the plans for the available resources will be generated and
represented in the model.

Each depot $d \in D$ has an inventory level modeled by variable z_{dpt}. An inventory
cost ϕ_{dp} is considered for each unit of product $p \in P$ in inventory at depot $d \in D$

between two subsequent periods. For a depot d, the inventory level of product p in time period t must not fall below a lower bound $\underline{z}_{dpt}$ (which typically corresponds to a safety inventory level) and must not exceed an upper bound (typically a maximum capacity) of $\overline{z}_{dpt}$ units. To allow violating the safety inventory and maximum allowed inventory levels, we let variable u_{dpt} represent how much the inventory level of product p falls below the safety inventory level at depot d in period t, q_{dpt} how much it exceeds the maximum allowed inventory level, and M^u_{dpt} and M^q_{dpt} corresponding penalty costs for violating the inventory constraints. It is assumed that, if any of the u:s or q:s are greater than zero in a period (i.e., an inventory constraint has been violated), a penalty is applied for resolving the inventory level infeasibility. There are other ways to model violation of inventory constraints, e.g., to include the u:s and q:s in constraint set (4) instead of in constraint sets (2) and (3). However, it is not obvious if one of these approaches is better than the other.

Binary decision variables are used to determine which transportation plans and production plans to use, and for obvious reasons exactly one transportation plan for each vehicle and exactly one production plan for each production line are allowed. Decision variable v_r determines if transportation plan $r \in R_v$ is used ($v_r = 1$) or not ($v_r = 0$), and w_s if production plan $s \in S_l$ is used ($w_s = 1$) or not ($w_s = 0$). We formulate our main problem (IMP) as the following MILP (MP is the LP-relaxation of IMP):

$$
\min \quad \sum_{p \in P} \sum_{d \in D} \sum_{t \in T} \left(\phi_{dp} z_{dpt} + M^q_{dpt} q_{dpt} + M^u_{dpt} u_{dpt} \right) +
$$
$$
\sum_{v \in V} \sum_{r \in R_v} \psi_r v_r + \sum_{l \in L} \sum_{s \in S_l} \omega_s w_s, \tag{1}
$$

$$
\text{s.t.} \quad z_{dpt-1} + \sum_{v \in V} \sum_{r \in R_v} v_r x_{dptr} + u_{dpt} - q_{dpt} - \varrho_{dpt} = z_{dpt},
$$
$$
t \in T, d \in D^C, p \in P, \tag{2}
$$

$$
z_{dpt-1} - \sum_{v \in V} \sum_{r \in R_v} v_r x_{dptr} + \sum_{l \in L} \sum_{s \in S_l} w_s y_{dpts} + u_{dpt} - q_{dpt} = z_{dpt},
$$
$$
t \in T, d \in D^P, p \in P, \tag{3}
$$

$$
\underline{z}_{dpt} \leq z_{dpt} \leq \overline{z}_{dpt}, \qquad\qquad d \in D, p \in P, t \in T, \tag{4}
$$

$$
\sum_{r \in R_v} v_r = 1, \qquad\qquad v \in V, \tag{5}
$$

$$
\sum_{s \in S_l} w_s = 1, \qquad\qquad l \in L, \tag{6}
$$

$$
v_r \in \{0,1\}, \qquad\qquad v \in V, r \in R_v, \tag{7}
$$

$$
w_s \in \{0,1\}, \qquad\qquad l \in L, s \in S_l, \tag{8}
$$

$$
z_{dpt}, u_{dpt}, q_{dpt} \in \mathbb{R}^+, \qquad\qquad d \in D, p \in P, t \in T. \tag{9}
$$

In IMP, the first component (i.e., the triple sum) of the objective function (1) models the inventory costs with penalties for violating the inventory constraints. The second component of (1) represents the cost for the transportation plans, and the third component the cost for production plans. Constraint sets (2) and (3) express the customer and producer depot inventory balances. For a customer depot, an inventory level at the end of a period equals the inventory level at the end of the previous period, plus the deliveries minus the consumption in the current period. For a producer depot, an inventory level at the end of a period equals the inventory level in the previous period, minus the pickups plus the produced amount in the current period. Constraints (4) assure that the inventory levels are kept between the minimum and maximum allowed levels, and (5) and (6) that exactly one plan is used for each resource (vehicle or production line).

For MP (as well as for RMP, which will be introduced below), we let λ denote the dual variables for constraints (2), μ the dual variables for constraints (3), δ the dual variables for constraints (5), and θ the dual variables for constraints (6). A dual variable expresses the change of the optimal objective function value per unit increase of the right hand side of the corresponding constraint. For example, $\bar{\lambda}_{dpt} > 0$ represents the value of having one extra unit of product p in depot d in period t, and a positive dual variable $\bar{\mu}_{dpt} > 0$ the value of having one extra unit of product p at depot d in period t. A value $\bar{\mu}_{dpt} < 0$ on the other hand, means that we want to decrease the production or reduce the inventory by transporting products away from depot d.

As mentioned above, to solve the IMP we relax the integer constraints and get an MP, i.e., a master problem. Furthermore, we only consider a subset of the potentially huge number of production and transportation plans, which gives an RMP, i.e., a restricted master problem. At initiation, RMP contains only a small number of plans for each resource (vehicle or production line), typically the procedure starts with an empty plan for each resource. For instance, for a vehicle an empty plan is one that has no deliveries and pickups. It should be emphasized that, to be able to satisfy constraints (5) and (6), RMP needs to contain at least one plan for each resource. We let $R'_v \subseteq R_v$ denote a set of all currently known transportation plans for vehicle $v \in V$, and $S'_l \subseteq S_l$ a set of currently known production plans for production line $l \in L$. That is, RMP is identical to MP, except for that we replace all occurrences of R_v with R'_v, all occurrences of S_l with S'_l.

In the solution approach, RMP is iteratively updated with new improving production and transportation plans that are generated by the planner agents based on the current optimal values of the λ, μ, δ, and θ dual variables. Dual variables are obtained from the current optimal solution (to RMP) when using a standard solver, which is typically based on the simplex method. A flow chart describing the main algorithm used by the coordinator agent is presented in Fig. 3, and a diagram showing how the coordinator agent communicates with the planners is provided in Fig. 4. In a distributed approach it is possible, and in many cases preferable, to send the transportation and production requests in parallel to allow the planners to solve subproblems simultaneously. Plans with negative reduced cost are added as improving columns/variables to RMP, and when no improving plan can be generated, the optimal solution to MP is obtained. Since the optimal solution to MP typically contains fractional combinations of plans, it is typically infeasible (due to the integer properties) in the original problem IMP. One approach for finding an optimal, or at least a heuristically "good" integer solution to IMP, some delivery/pickup (depot, period, product, and vehicle) or production (period, product and production line) must be restricted (fixed) to some integer quantity whenever a fractional optimal solution to RMP is obtained. To handle the integer characteristics of the studied problem, we apply branch-and-price [10], however in a limited form. To guarantee that the optimal solution to IMP will be generated, it is necessary to use a search tree, in which each node represents a restricted version of the original MP due to additional restrictions. In the branching approach, all possible combinations of fixings are represented somewhere in the search tree, and in theory the optimal solution of the original problem is guaranteed. However, since time and memory are limited, this is not possible for the studied problem. Due to its characteristics of being NP-hard and the size of real-world scenarios, we choose to explore only one branch in the search tree. Variable restrictions are determined by the coordinator and communicated to the planners to prevent them from generating future plans that violate variable fixings. Throughout the procedure of the algorithm, more and more variables are fixed, and eventually the algorithm terminates with an integer solution to IMP. In Section 5.1 we give a more detailed account to our strategy for variable fixings and termination.

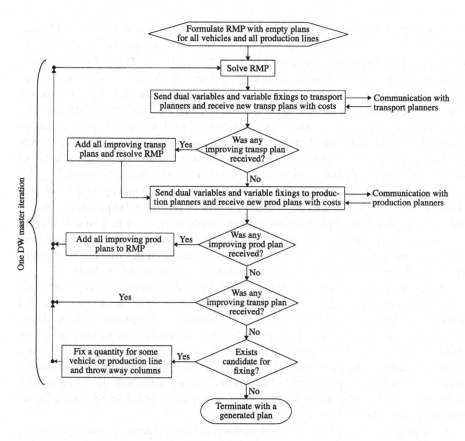

Fig. 3. A flow chart of the main algorithm used by the supply chain coordinator agent

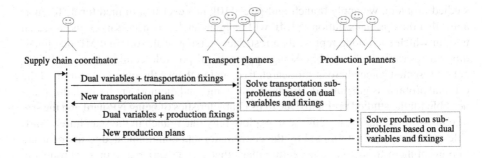

Fig. 4. Diagram describing the communication between the coordinator and planner agents

5.1 Heuristic Strategy for Variable Fixing and Termination

In our variable fixing strategy, which is inspired by the approach by Persson and Göete-Lundgren [21], we let $f_{ptl}^{P} \in \mathbb{Z}^{+}, p \in P, t \in T, l \in L$ represent production fixings (i.e., minimum quantities to produce), and $f_{dptv}^{T} \in \mathbb{Z}^{+}, d \in D, p \in P, t \in T, v \in V$ transportation fixings (i.e., minimum quantities to deliver and pickup). That is, there is one (production) fixing parameter for each combination of product, period and production line, and one (transportation) fixing parameter for each combination of depot, product, period and vehicle. All f^{P}:s and f^{T}:s are initialized to 0, meaning that no quantities are fixed when the algorithm starts. A fixing of a quantity f_{ptl}^{P} means that all subsequent production plans that are generated for production line l must contain a production of at least f_{ptl}^{P} units of product p in period t. Similarly, a transportation fixing f_{dptv}^{T} for vehicle v means that all subsequent plans for v must contain a pickup (if $d \in D^{P}$) or delivery (if $d \in D^{C}$) of at least f_{dptv}^{T} units of product p for depot d in period t. Hence, it follows that already fixed productions, pickups and deliveries may be re-fixed to higher values later in the process, as long as the capacities of vehicles and production lines are met when new plans are created. After a fixing has been determined, all columns (plans) violating the fixing (i.e., infeasible columns) are removed from RMP.

The main idea in the variable fixing strategy is that the production, pickup or delivery with the highest representation in the optimal solution to RMP should be chosen for fixing. For instance, to calculate how much a production is represented in RMP, the values representing how much the plans are used in the solution are added together for those plans (for the particular production line) that have a production strictly greater than zero units for the particular product and period. The calculation of how much a pickup or delivery is represented in RMP is made in the same way. Hence, a number between 0 and 1 is obtained for each production, pickup and delivery, and a higher number means a higher representation in RMP.

For transportation, we find out which pickup or delivery (depot, product, period and vehicle) that is most represented in RMP by calculating

$$(d', p', t', v') = \operatorname*{arg\,max}_{d \in D, p \in P, t \in T, v \in V} \sum_{r \in R'_v} v_r \Lambda_{dptr}^{T}, \tag{10}$$

where

$$\Lambda_{dptr}^{T} = \begin{cases} 0 & \text{if } x_{dptr} = 0 \\ 1 & \text{if } x_{dptr} \neq 0. \end{cases} \tag{11}$$

To avoid fixing pickups and deliveries that are represented in all columns in RMP, and to disregard those pickups and deliveries that are not represented at all, in equation (10) we require that

$$0 < \sum_{r \in R'_{v'}} v_r \Lambda_{d'p't'r}^{T} < 1. \tag{12}$$

Since it is only relevant to consider pickups and deliveries that are represented by a strictly higher quantity than previously fixed, we also require that

$$\left| \sum_{r \in R'_{v'}} v_r x_{d'p't'r} \right| > f_{d'p't'v'}^{T}. \tag{13}$$

In the same way as for transportation, we find out which production (production line, product and period) is most represented in RMP by calculating

$$(l', p', t') = \arg\max_{l \in L, p \in P, t \in T} \sum_{s \in S'_l} w_s \Lambda_{d(s)pts}^{P} \tag{14}$$

where

$$\Lambda_{dpts}^{P} = \begin{cases} 0 & \text{if } y_{dpts} = 0 \\ 1 & \text{if } y_{dpts} \neq 0, \end{cases} \tag{15}$$

and where it is required in equation (14) that

$$0 < \sum_{s \in S'_l} w_s \Lambda_{d(s)p't's}^{P} < 1, \tag{16}$$

and

$$\sum_{s \in S'_{l'}} w_s y_{d(s)p't's} > f_{l'p't'}^{P}, \tag{17}$$

If only (d', p', t', v') exists (i.e., no candidate for production fixing exists), or if

$$\sum_{r \in R'_{v'}} v_r \Lambda_{d'p't'r}^{T} \geq \sum_{s \in S'_{l'}} w_s \Lambda_{d(s)p't's}^{P}, \tag{18}$$

then $f_{d'p't'v'}^{T}$ (i.e., a transportation fixing) is fixed (or re-fixed) for vehicle v' according to

$$f_{d'p't'v'}^{T} = \left\lceil \left| \sum_{r \in R'_{v'}} v_r x_{d'p't'r} \right| \right\rceil. \tag{19}$$

Otherwise, if only (l', p', t') exists, or if

$$\sum_{r \in R'_{v'}} v_r \Lambda_{d'p't'r}^{T} < \sum_{s \in S'_{l'}} w_s \Lambda_{d(s)p't's}^{P}, \tag{20}$$

then $f^P_{l'p't'}$ (i.e., a production fixing) will be fixed (or re-fixed) for production line l' according to

$$f^P_{l'p't'} = \left\lceil \sum_{s \in S'_{l'}} w_s y_{d(s)p't's} \right\rceil. \tag{21}$$

If neither a pickup or delivery pickup fixing candidate (d', p', t', v'), or a production fixing candidate (l', p', t') can be found, there might still exist any two columns $r'', r''' \in R'_{v'}$ or $s'', s''' \in S'_{l'}$ (for a vehicle $v' \in V$ or a production line $l' \in L$) with different coefficients in the optimal solution to RMP. This scenario can occur if equation (12) or (16) equals one for some delivery, pickup or production while $x_{d'p't'r''} \neq x_{d'p't'r'''}$ or $y_{d'p't's''} \neq y_{d'p't's'''}$ for any $d' \in D, p' \in P$ and $t' \in T$. Then we either set

$$f^T_{d'p't'v'} = \left\lceil \left| \sum_{r \in R'_{v'}} v_r x_{d'p't'r} \right| \right\rceil \tag{22}$$

or

$$f^P_{l'p't'} = \left\lceil \sum_{s \in S'_{l'}} w_s y_{d(s)p't's} \right\rceil, \tag{23}$$

to be able to converge towards an integer solution to IMP. Moreover, we remove from RMP all columns $r \in R'_{v'}$ with parameter

$$|x_{d'p't'r}| \neq f^T_{d'p't'v'} \tag{24}$$

if a delivery or a pickup was fixed, or all columns $s \in S'_{l'}$ with parameter

$$y_{d(s)p't's} \neq f^P_{l'p't'} \tag{25}$$

if a production was fixed.

From a few small-scale experiments, we realized that the convergence rate of the column generation approach was too slow to allow the RMPs to be solved to optimality before considering termination or variable fixing. Instead we use a heuristic termination strategy that is based on the relative improvement of generated plans, resulting in heuristic solutions to the RMPs (and hence MPs). The idea is that it still will be possible to find heuristically good integer solutions to IMP at termination of the algorithm.

A solution to RMP is considered to be good enough when: for each resource, the average reduced cost of the e (actually e^V for a vehicle or e^P for a production line) most recently added plans (for the particular resource) is less than g (g^V or g^P) percent better than the average reduced cost of the e (e^V or e^P) plans that were added immediately before that. That is, plans are added until the improvement rate has decreased to a certain level. At this point the algorithm either determines a variable to fix if the current solution of RMP is fractional, or terminates if the solution is integer. The choice of e^V, e^P, g^V, and g^P is typically a trade-off between solution time and quality.

5.2 Transportation Subproblems

We formulate a transportation subproblem for a vehicle $v \in V$ using a hierarchical approach with two separated subproblems; one routing problem and one product assignment problem. The routing problem is formulated as a shortest path problem with additional constraints for representing pickup and delivery fixings. The output is a sequence of depots with corresponding time-periods, which will be visited by the vehicle throughout the planning horizon. In other words, the output is a route serving as input to the product assignment problem. Then the product assignment problem decides how much of each product should be delivered for each customer depot visit in the route.

For the studied problem it was found reasonable to allow a maximum of two customer depot visits before visiting a producer. Formally this assumption can be described by introducing a set of network layers, denoted $H = \{0, \ldots, \bar{h}\}$, as a means to prevent vehicles from visiting more than $\bar{h}$ customers before visiting some producer. In our problem formulation $\bar{h} = 2$, and this is the largest number that makes it possible to calculate analytically correct dual variables for the routing subproblems.

The idea is that layer 0 belongs to the producer depots and layer 1 through $\bar{h}$ to the customer depots (each customer is represented in all layers from 1 to $\bar{h}$). A transport to a customer arrives in one layer higher than the departure layer, and a transport to a producer always arrives in layer 0 regardless of which is the departure layer. As an example, a transport from a producer depot to a customer depot starts in layer 0 and ends in layer 1. Accordingly, after $\bar{h}$ customer depot visits it is only possible to travel to a producer depot. However, it is possible to return to a producer before $\bar{h}$ customer depots have been visited. An *outbound trip* for a vehicle v is defined as a trip that starts at a producer depot, visits a number of customer depots, and ends when v returns to some producer. We require that an outbound trip must contain at least one customer, which is why a transport between two producers is not considered as an actual outbound trip even though it typically is allowed to travel directly between producers. The optimal route from the routing subproblem is defined as a set of outbound trips, which we denote O_v. A small example, including only a few transport options, of a time expanded transportation network is shown in Fig. 5. In the example, 3 network layers are used to allow a maximum of 2 customer depot visits before visiting a producer.

A time expanded transportation network for vehicle v is defined as a directed graph $(\mathcal{N}_v \cup \{a\}, \mathcal{A}_v)$ with a set of nodes $\mathcal{N}_v \cup \{a\}$ and a set of arcs $\mathcal{A}_v$. The set $\mathcal{N}_v = \{n_{dht} : d \in D, h \in H, t \in T\}$ is a set of network nodes corresponding to actual

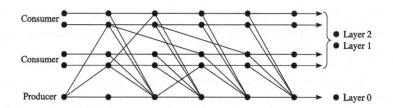

Fig. 5. An example of a time expanded transportation network, in which a maximum of two customer depot visits are allowed before returning to the producer depot

depots, and a is an artificial node that allows v to end its route anywhere in the network. An arc $(n_i, n_j) \in \mathcal{A}_v$, with travel cost $c^{RP_v}_{(n_i, n_j)}$, is a connection between a starting node $n_i \in \mathcal{N}_v$ and an ending node $n_j \in \mathcal{N}_v \cup \{a\}$. We introduce parameter $\eta_{(n_i, n_j)}$ as a function of the actual dual variables μ and λ to describe an extra cost or discount that is added to the cost of arc (n_i, n_j). The calculation of the $\eta_{(n_i, n_j)}$ parameters is made in a way that the structure of the transportation network is being utilized, as will be detailed below. We introduce decision variable $x^{RP_v}_{(n_i, n_j)} \in \{0, 1\}$ to determine the usage of arc $(n_i, n_j) \in \mathcal{A}_v$ so that $x^{RP_v}_{(n_i, n_j)} = 1$ if arc (n_i, n_j) is used in the solution, otherwise $x^{RP_v}_{(n_i, n_j)} = 0$.

The routing subproblem RP_v for vehicle v is formulated as a standard minimum cost flow problem as:

$$\min \sum_{(n_i, n_j) \in \mathcal{A}_v} (c^{RP_v}_{(n_i, n_j)} - \eta_{(n_i n_j)}) x^{RP_v}_{(n_i, n_j)} \tag{26}$$

$$\text{s.t.} \sum_{n_k:(n_k, n_i) \in \mathcal{A}_v} x^{RP_v}_{(n_k, n_i)} - \sum_{n_j:(n_i, n_j) \in \mathcal{A}_v} x^{RP_v}_{(n_i, n_j)} = b_{n_i}, \quad n_i \in \mathcal{N}_v \cup \{a\}, \tag{27}$$

$$x^{RP_v}_{(n_i, n_j)} \in \{0, 1\}, \qquad (n_i, n_j) \in \mathcal{A}_v.$$

Constraint set (27) specifies the node balance constraints where b_{n_i} follows the rules in equation (28). Node $n_s \in \mathcal{N}_v$ denotes the node where v is situated at the beginning of the planning period, and a allows v to be at any node at the end of the planning period. The node balance parameter b_{n_i} for node n_i is defined as

$$b_{n_i} = \begin{cases} -1 & \text{if } n_i = n_s \\ 1 & \text{if } n_i = a \\ 0 & \text{otherwise.} \end{cases} \tag{28}$$

The arc set $\mathcal{A}_v$ can be described as a subset of the union

$$\mathcal{A}^{WP}_v \bigcup \mathcal{A}^{PP}_v \bigcup \mathcal{A}^{PC}_v \bigcup \mathcal{A}^{CC}_v \bigcup \mathcal{A}^{CP}_v \bigcup \mathcal{A}^{A}_v,$$

where the content of each of these sets will be detailed below. The estimated time $t^{link}_{d'd''hv}$, that vehicle v needs for traveling the direct link from depot d' to depot d'', starting in layer h, includes actual driving time, estimated times for loading and unloading, and estimated resting time. How to calculate link-traveling times will be detailed below. We here represent a network node using three indices; depot, layer and time period.

$\mathcal{A}^{WP}_v = \{(n_{d0t}, n_{d0,t+1}) : d \in D^P, t \in T \setminus \{\bar{t}\}\}$ contains arcs going from a producer to the same producer. This allows v to wait at a producer depot between to subsequent time periods.

$\mathcal{A}^{PP}_v = \{(n_{d'0t}, n_{d''0,t+\lceil t^{link}_{d'd''0v}\rceil}) : d', d'' \in D^P, d' \neq d'', t \in \{0, \ldots, \bar{t} - t^{link}_{d'd''0v}\}\}$ contains arcs going from one producer to a different producer to allow v to travel between producer depots.

$\mathcal{A}^{PC}_v = \{(n_{d'0t}, n_{d''1,t+\lceil t^{link}_{d'd''0v}\rceil}) : d' \in D^P, d'' \in D^C, t \in \{0, \ldots, \bar{t} - t^{link}_{d'd''0v}\}\}$ contains arcs going from a producer to a customer, allowing v to travel from producer depots to customer depots.

$\mathcal{A}_v^{CC} = \{(n_{d'ht}, n_{d'',h+1,t+\lceil t^{\text{link}}_{d'd''hv}\rceil}) : d', d'' \in D^C, d' \neq d'', h \in \{1, \ldots, \bar{h} - 1\}, t \in \{0, \ldots, \bar{t} - t^{\text{link}}_{d'd''hv}\}\}$ contains arcs going from a customer to a different customer allowing v to travel between customer depots.

$\mathcal{A}_v^{CP} = \{(n_{d'ht}, n_{d''0,t+\lceil t^{\text{link}}_{d'd''hv}\rceil}) : d' \in D^C, d'' \in D^P, h \in H \setminus \{0\}, t \in \{0, \ldots, \bar{t} - t^{\text{link}}_{d'd''hv}\}\}$ contains arcs going from a customer to a producer to allow v to travel from customer depots to producer depots.

$\mathcal{A}_v^A = \{(n, a) : n \in \mathcal{N}_v\}$ contains one arc from each network node to the artificial node a to allow v to stop its route at any location, in any layer, and in any time period. The transportation cost for an "artificial arc", starting in node n, corresponds to the cost for traveling from n to the "home base" of v. The reason for adding costs to the artificial arcs, even though they do not correspond to actual transports, is that we do not want any particular location to be favored at the end of the planning period.

If there is no direct connection between two depots, $\mathcal{A}_v$ will not contain any arcs between the corresponding network nodes. Moreover, the time expanded transportation networks contain no arcs allowing vehicles to wait at customer depots. We consider this modeling assumption reasonable, because in the considered problem there is no need for waiting at customers.

Transportation costs for arcs in the routing subproblems are composed of three types of costs, which in the model are represented by link-based costs:

1. Time-based costs (e.g., driver salary, capital cost, and administration) are assumed for the time the vehicle spends away from a producer depot. The driver is assumed to receive salary when the vehicle is on the road and during unloading of products. Loading of products is performed by ground staff, who has the same salary as the drivers. Unloading, on the other hand, is assumed to be performed by the drivers. Therefore, the driving time, as well as the time for unloading need to be compensated for by resting time, and salary is not considered during resting.
2. Distance-based costs (e.g., fuel, vehicle wear, and kilometer taxes) are based on the distance the vehicles travel.
3. Link-based costs (e.g., road tolls) are fixed costs that are charged when vehicles travel on certain links.

When a routing subproblem is formulated there exists no information about loadings and unloadings (e.g., concerning quantities). Therefore, loading and unloading times have to be estimated, and we have chosen to base these estimations on average loading and unloading times taken over all available product types. We assume loading times for a full vehicle at producer nodes, and unloading times for a 50% load at customer depots, i.e., loading time is added to arcs leaving producers and unloading time to arcs arriving to customer depots. Accordingly, we model fixed times for loading and unloading, which are independent on actual transportation volumes.

The working and resting hour regulation used in the routing subproblem allows for a maximum of $t^{\text{max working}}$ hours of working (driving) before there must be a minimum of $t^{\text{min resting}}$ hours of resting. However, we assume that two outbound trips that in sequence violate the working hour regulation are performed by different drivers. Hence, working hour restrictions are only considered within a single outbound trip.

To be able to estimate the time that the driver needs to rest when traveling on a link from depot d' to depot d'', starting in layer h, we introduce a working hour estimation $t_{d'hv}^{\text{est working}}$. This is a lower bound estimation of the time it takes for v to drive from any producer depot to the starting depot d', with the requirement that it must pass exactly h different customer depots. The estimation is actually the shortest path from any producer to d' with the additional requirement of exactly h customer depot visits. For small networks, and small numbers of $\bar{h}$, these lower bound estimations can be found rather easily.

The estimated time $t_{d'd''hv}^{\text{link}}$ that vehicle v needs to travel a link from depot d' to depot d'', when starting in layer h, is calculated according to equation

$$t_{d'd''hv}^{\text{link}} = t_{d'd''hv}^{\text{working}} + t_{d'd''hv}^{\text{resting}}, \tag{29}$$

where the link traveling time is decomposed into working time $t_{d'd''hv}^{\text{working}}$ (i.e., time for driving and unloading) and resting time $t_{d'd''hv}^{\text{resting}}$. The resting time $t_{d'd''hv}^{\text{resting}}$ is estimated according to equation

$$t_{d'd''hv}^{\text{resting}} = k^d \cdot t^{\text{min resting}}, \tag{30}$$

as the number of resting periods k^d times the minimum resting time $t^{\text{min resting}}$ of each such resting period. The number of resting periods depends on the link working time $t_{d'd''hv}^{\text{working}}$ and on the working time estimation $t_{d'hv}^{\text{est working}}$ for traveling to depot d', and it is calculated as

$$k^d = \max \left\{ \left\lfloor \frac{\left(t_{d'hv}^{\text{est working}} - t^{\text{est load}} \right) \bmod t^{\text{max working}} + t_{d'd''v}^{\text{working}} - \varepsilon}{t^{\text{max working}}} \right\rfloor, 0 \right\}, \tag{31}$$

where the expression

$$\left(t_{d'hv}^{\text{est working}} - t^{\text{est load}} \right) \bmod t^{\text{max working}} \tag{32}$$

denotes the remaining portion of the estimated working time $t_{d'hv}^{\text{est working}}$ that has not been accounted for by resting time when traveling to d'. The loading time is subtracted from the estimated working time $t_{d'hv}^{\text{est working}}$ because loading is assumed to be performed by ground staff. The parameter $0 < \varepsilon < 1/t^{\text{max working}}$ is introduced in order to avoid a special case that occurs in equation (31) whenever the denominator is a devisor of the nominator, causing k^d to take on a value that is 1 unit too large. It is worth noting that all times are expressed in minutes represented by integer numbers. It should also be noted that the estimation of resting times assumes that vehicles never wait at customer depots, which is also prohibited in the routing problem.

As mentioned above, the η cost parameters that are used in the routing subproblems are calculated from the μ and λ dual variables in a way that the structure of the transportation network is utilized. For each arrival customer depot, network layer and time period, the calculation of an analytically correct value of the η parameter requires knowledge about when and from which producer depot the vehicle has departed. To enable correct costs and strong estimates on time, we make the following assumptions:

1. An outbound trip is allowed to contain at most two customer depot visits.
2. Each customer receives products from exactly one producer depot.
3. Waiting at customer depots is not allowed.

With these assumptions, the estimated resting times will also be analytically correct, i.e., not estimated. Otherwise it is impossible to know how vehicles travel in the network. The assumptions listed above are essential because driving time, which has not been "rested for" accumulates over traveled links. A small example of a driving and resting time approximation is given in Fig. 6.

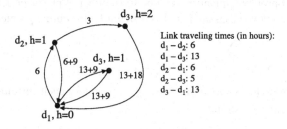

Fig. 6. An example of a working and resting time approximation with $t^{\text{max working}} = 10$ and $t^{\text{min resting}} = 9$

From the maximum capacity of vehicle v (weight capacity φ_v^{weight} and volume capacity φ_v^{volume}), for each $p \in P$ we estimate the maximum number

$$\Phi_{vp} = \left\lfloor \min \left\{ \frac{\varphi_v^{weight}}{weight(p)}, \frac{\varphi_v^{volume}}{volume(p)} \right\} \right\rfloor \quad (33)$$

of items of product p that can be loaded on v. The Φ_{vp} :s are restricted either by the volume capacity or by the weight capacity of v, and they will be used when calculating the η parameters, as well as in the product assignment problem presented below.

In order to compute the parameter η we first consider transports from producer depots to customer depots. A transport is assumed to start in a network node n_s representing producer depot $d(n_s) \in D^P$ and period $t(n_s)$, and it is assumed to end in a network node n_e representing customer depot $d(n_e) \in D^C$ and period $t(n_e)$. This case is rather straightforward and the $\eta_{(n_s,n_e)}$ value is calculated according to

$$\eta_{(n_s,n_e)} = \max \left\{ 0, \max_{p \in P'} \left((\lambda_{d(n_e)pt(n_e)} - \mu_{d(n_s)pt(n_s)}) \cdot \Phi_{vp} \right) \right\}, \quad (34)$$

where $P' \subseteq P$ describes the set of products that can be produced in $d(n_s)$ and consumed in $d(n_e)$. Next we consider transports between two different customer depots. Here a transport is assumed to start in a network node n_e representing customer depot

$d(n_e) \in D^C$ and period $t(n_e)$, and it is assumed to end in a network node n_c representing customer depot $d(n_c) \in D^C$ and period $t(n_c)$. Knowledge about the network node n_s representing the producer depot $d(n_s) \in D^P$ and the period $t(n_s)$ from where the transport was assumed to start is required for a correct calculation of $\eta_{(n_e, n_c)}$. As mentioned above, such information is accessible since: at most two customer visits is allowed in an outbound trip, each customer can be reached from at most one producer and waiting at customer depots is forbidden. The value of $\eta_{(n_e, n_c)}$ can be calculated as

$$\eta_{(n_e, n_c)} = \max\left\{0, \eta_{(n_s, n_c)} - \eta_{(n_s, n_e)}\right\}, \tag{35}$$

where $\eta_{(n_s, n_e)}$ and $\eta_{(n_s, n_c)}$ is calculated according to equation (34). Hence the value for going from n_e to a customer depot n_c is equivalent to the potential extra value that can be obtained at n_c compared to the value at n_e.

In a routing subproblem, quantities can be fixed either at producer depots or at customer depots and the two cases are handled differently. In routing problem RP_v, a constraint

$$\sum_{n_i : d(n_i) = d', t(n_i) = t'} \sum_{n_k : (n_i, n_k) \in \mathcal{A}_{v'}} x^{\text{RP}_{v'}}_{(n_i, n_k)} = 1 \tag{36}$$

is added for each producer depot fixing $f^T_{d'p't'v'} > 0$, and for each customer depot fixing $f^T_{d'p't'v'} > 0$ we add a constraint

$$\sum_{n_k : (n_k, n_j) \in \mathcal{A}_{v'}} \sum_{n_j : d(n_j) = d', t(n_j) = t'} x^{\text{RP}_{v'}}_{(n_k, n_j)} = 1. \tag{37}$$

Constraint (36) means that v' must depart from producer depot d' in period t', and constraint (37) forces vehicle v' to arrive at customer depot d' in period t'. It should be emphasized that these constraints remove the integrality property of the routing problems.

Implicitly the product assignment subproblem is solved already in the routing subproblem. However, we explicitly formulate an optimization problem in order to handle fixings, i.e., minimum quantities for pickups and deliveries. From the optimal route O_v determined by RP_v, defined as a set of outbound trips, a product assignment problem is formulated. The purpose of a product assignment problem is to decide how much of each product will be picked-up from each producer depot and delivered to each customer depot in the route. The problem separates into one subproblem (ASP_o) for each outbound trip $o \in O_v$.

We introduce an ordered index set $J_o = \{1, \dots, j_o\}$ over the depot visits in outbound trip $o \in O_v$, starting with index 1 for the producer depot, index 2 for the first customer depot, etc. For simplified representation we let $d(j)$ and $t(j)$ refer to the depot and time period represented by visit j in the outbound trip. Moreover, we let decision variable $x^{\text{ASP}_o}_{d(j)p} \in \mathbb{Z}^+, p \in P, j \in J_o \setminus \{1\}$ represent the amount of product type p that is delivered to the j:th customer in outbound trip o.

We formulate the product assignment problem ASP_o for outbound trip $o \in O_v$ as:

$$\max \quad \sum_{p \in P} \sum_{j \in J_o \setminus \{1\}} \left(\lambda_{d(j)pt(j)} - \mu_{d(1)pt(1)} \right) x_{d(j)p}^{\mathrm{ASP}_o}, \tag{38}$$

$$\text{s.t.} \quad \sum_{p \in P} \sum_{j \in J_o \setminus \{1\}} weight(p) \cdot x_{d(j)p}^{\mathrm{ASP}_o} \leq \varphi_v^{weight}, \tag{39}$$

$$\sum_{p \in P} \sum_{j \in J_o \setminus \{1\}} volume(p) \cdot x_{d(j)p}^{\mathrm{ASP}_o} \leq \varphi_v^{volume}, \tag{40}$$

$$x_{d(j)p}^{\mathrm{ASP}_o} \in \mathbb{Z}^+, \qquad\qquad p \in P, j \in J_o \setminus \{1\}. $$

The objective function (38) maximizes the utility of pickup and deliveries for the given route, and constraint sets (39) and (40) express the weight and volume restrictions on vehicle v.

In a product assignment problem, a quantity can be fixed either for a producer depot or for a customer depot. For each producer depot fixing $f_{d'p't'v'}^T > 0$, a constraint

$$\sum_{j \in J_{o'} \setminus \{1\}} x_{d(j)p'}^{\mathrm{ASP}_{o'}} \geq f_{d'p't'v'}^T \tag{41}$$

is added to the assignment problem $\mathrm{ASP}_{o'}$ that represents $f_{d'p't'v'}^T$. This forces v' to pickup at least $f_{d'p't'v'}^T$ units of product p' from producer depot d' in period t'. Moreover, for each customer fixing $f_{d'p't'v'}^T > 0$, we add a constraint

$$x_{d'p'}^{\mathrm{ASP}_{o'}} \geq f_{d'p't'v'}^T, \tag{42}$$

to the assignment problem $\mathrm{ASP}_{o'}$ that represents $f_{d'p't'v'}^T$. This guarantees that at least $f_{d'p't'v'}^T$ units of product p' will be delivered to customer depot d' in period t'.

The ASP_o subproblems assign products to the optimal route O_v, and together they form a transportation plan. After solving the routing subproblem and the product assignment subproblems, the optimal objective function value (i.e., the reduced cost in RMP) of the transportation subproblem for vehicle v can be calculated as

$$\sum_{(n_i, n_j) \in \mathcal{A}_v} c_{(n_i, n_j)}^{\mathrm{RP}_v} x_{(n_i, n_j)}^{*\mathrm{RP}_v} + \sum_{o \in O_v} \mathrm{ASP}_o^* - \delta_v, \tag{43}$$

where $x_{(n_i, n_j)}^{*\mathrm{RP}_v}$ denotes the optimal value of variable $x_{(n_i, n_j)}^{\mathrm{RP}_v}$, ASP_o^* denotes the optimal objective value of ASP_o, and δ_v is the convexity constraint dual variable for vehicle v.

5.3 Production Scheduling Subproblems

The purpose of a production scheduling subproblem for a production line $l \in L$ is to find improving production plans for l guided by the values of the μ dual variables. We let c_{lp}^{prod} denote the cost for production line l to produce one unit of product $p \in P$, and c_{lp}^{setup} denote a fixed setup cost for each period product p is produced. The modeled real-world problem contains costs for startup and changeover, but we consider the problem formulation here to be detailed enough. A more advanced model for the studied real-world production problem is provided by Sohier [25]. The model by Sohier includes product sequencing on a set of production lines, but it is restricted to short planning horizons, which makes it difficult for us to use.

We let decision variable $x_{pt}^{\text{PSP}_l} \in \mathbb{Z}^+$ determine the amount of product p to be produced in period t. The products that are produced in a period are assumed to be available for pickup in the same period. Binary variable $y_{pt}^{\text{PSP}_l} \in \{0,1\}$ is used to indicate whether there will be a production of product p in period t ($y_{pt}^{\text{PSP}_l} = 1$) or not ($y_{pt}^{\text{PSP}_l} = 0$), and we let U denote the maximum number of different product types that can be produced in one period. The production scheduling subproblem PSP$_l$ can be formulated as:

$$\min \quad \sum_{p \in P} \sum_{t \in T} \left(c_{lp}^{\text{prod}} - \mu_{pt} \right) x_{pt}^{\text{PSP}_l} + \sum_{p \in P} \sum_{t \in T} c_{lp}^{\text{prod}} y_{pt}^{\text{PSP}_l} - \theta_l \tag{44}$$

$$\text{s.t.} \quad \sum_{p \in P} t_{lp}^{\text{prod}} x_{pt}^{\text{PSP}_l} \leq \tau, \qquad\qquad t \in T, \tag{45}$$

$$t_{lp}^{\text{prod}} x_{pt}^{\text{PSP}_l} \leq \tau y_{pt}^{\text{PSP}_l}, \qquad\qquad p \in P, t \in T, \tag{46}$$

$$\sum_{p \in P} y_{pt}^{\text{PSP}_l} \leq U, \qquad\qquad t \in T, \tag{47}$$

$$x_{pt}^{\text{PSP}_l} \in \mathbb{Z}^+, \qquad\qquad p \in P, t \in T,$$

$$y_{pt}^{\text{PSP}_l} \in \{0,1\}, \qquad\qquad p \in P, t \in T.$$

Constraints (45) models the capacity constraints, where t_{lp}^{prod} denotes the time needed for production line l to produce one unit of product p, and τ denotes the length of a time period. To be able to model setup costs, constraint set (46) forces each $y_{pt}^{\text{PSP}_l}$ variable to value one whenever the corresponding $x_{pt}^{\text{PSP}_l}$ is greater than 0. Constraint set (47) restricts the number of different product types can be produced in any period to U. Note that we subtract the convexity constraint dual variable θ_l in the objective function, and that the inventory balance constraints normally included in production scheduling problems here belong to the master problem, which is controlled by the supply chain coordinator. Therefore, the production scheduling problems separate over time. The objective function value of the optimal solution gives the reduced cost of the generated production plan.

To represent production fixings in the production scheduling subproblems, for each production fixing $f_{p't'l'}^{P} > 0$ we add a constraint

$$x_{p't'}^{\text{PSP}_{l'}} \geq f_{p't'l'} \tag{48}$$

to subproblem PSP$_{l'}$. This forces production line l' to include a production of at least $f_{p't'l'}$ units of product p' in period t' in all subsequently generated plans.

6   Computational Experiments

Our DW column generation algorithm has been implemented inside a multi-agent-based simulation tool called TAPAS [24]. TAPAS is implemented in the Java programming language using the Java Agent DEvelopment Framework (JADE) platform [26] and the MILP solver ILOG CPLEX[1] 10.0 is used for solving optimization problems. In the experiments TAPAS runs on a single computer, but a distributed implementation where agents run on different computers, is straightforward. This would increase the communication overhead while potentially reducing the overall solution time due to the availability of more processors and the potential for parallel processing. In our implementation, overhead is caused by JADE, e.g., by spending effort on coding and decoding messages. Our implementation approach simulates a situation with a coordinator agent and planning agents located at different locations, and it allows for investigation of the potential usage of the approach.

6.1 Scenario Description

The proposed solution approach has been used for solving 5 scenarios with a transportation network containing 2 producer depots d_1 and d_7 (with one production line each), 6 customer depots $d_2, \ldots, d_6$ and d_8, and a planning horizon of 72 time periods with uniform length 2 hours (i.e., 6 days). The scenarios were generated randomly, and they differ in customer demand, as well as minimum, maximum and initial storage levels. In each scenario, three different settings regarding VMI were tested, giving a total of 15 simulation runs. Producer depot d_1 provides customer depots $d_2, \ldots, d_6$ with products and the purpose of d_7 and d_8 is to model possibly non-empty return transports for a small portion of the vehicles returning to d_1. The routes between depots are represented by direct links, with distances given in Table 1. From depot d_1 it is possible to travel to depots $d_2, \ldots, d_7$, from $d_2, \ldots, d_6$ to all depots except d_8, from d_7 to d_1 and d_8, and from d_8 to d_1. The average speed is 70 km/hour on all links and the time-based transportation cost is estimated to 250 SEK/hour. Furthermore, the EEC regulations (EEC 3820/1985) for working and resting hours are approximated by allowing a maximum of 10 hours of working before a minimum of 9 hours of resting has to take place. Moreover, the distance-based cost during transportation is 7.69 SEK/km.

Table 1. Distance (in km) between the depots in the scenario

d_1							
355	d_2						
353	216	d_3					
450	407	167	d_4				
531	487	248	141	d_5			
615	455	333	226	93	d_6		
209	130	232	362	442	522	d_7	
95	-	-	-	-	-	119	d_8

The scenarios use a fleet of 9 vehicles $v_1, \ldots, v_9$ with identical weight capacities of 35 tons. The volume restriction for the vehicles are dominated by the weight restrictions,

---

[1] http://www.ilog.com/

and there are no limitations concerning which types of products can be transported on the different vehicles. For each vehicle, we specify a depot and an earliest time period from which it is available for further planning. Vehicles v_8 and v_9 represent third party transport capacity, which can be called in to a cost approximately 10% higher than the cost for using $v_1, \ldots, v_7$.

The scenarios use 4 different product types $p_1, \ldots, p_4$, each with a mass of 5 tons per unit (or batch). Product type p_4 is used to model return transports and it is assumed that p_4 is produced only in d_7 and consumed only in d_8. Therefore, the production and setup costs for p_4 are assumed to be 0. Products $p_1, \ldots, p_3$ can be produced in production line l_1 with production costs of 1500 SEK/unit and setup costs of 1000 SEK. The production capacity of product types $p_1, \ldots, p_3$ is 2.5 batches/hour and the capacity for producing p_4 is 4 batches/hour. The cost for storing one unit of $p_1, \ldots, p_3$ between two subsequent periods is 300 SEK in d_1 and 350 SEK in $d_2, \ldots, d_6$. For p_4, the storage cost equals 150 SEK in d_7 and 175 SEK in d_8.

For each producer depot, an initial inventory level was chosen randomly between 0 and a maximum inventory level, which is 10 for products $p_1, \ldots, p_3$ in d_1, and 7 for product p_4 in d_7. The penalty cost for violating the allowed minimum and maximum inventory levels is 8200 SEK/unit for d_1 and 4120 SEK/unit for d_7. For customer depots, the penalty cost for exceeding the maximum allowed inventory level is 8200 SEK/unit for depots $d_2, \ldots, d_6$ and 4120 SEK/unit for d_8. For inventory shortages, the penalty is 9020 SEK/unit for depots $d_2, \ldots, d_6$ and 4520 SEK/unit for depot d_8.

In depots $d_2, \ldots, d_6$, the forecasted average consumptions (in units per 2 hour time period) of products $p_1, \ldots, p_3$ is 0.15, and in d_8, the average forecasted consumption of p_4 is 0.35. A demand-forecast used in a scenario is chosen randomly in steps of 0.05 units with equal probability between 0 and 2 times the averaged forecasted demand. For example, for an expected forecast of 0.2, values between 0 and 0.4 can be generated. Consumption is aggregated to integer values, which means, for instance, that an average consumption of 0.2 units per period gives a demand for 1 unit every 5:th period. The main reason for aggregating consumptions is that transportation is performed in integer quantities, and it simplifies our comparisons between scenarios with and without VMI.

In a customer depot, the safety inventory level for a product is chosen as the quantity that, according to the consumption forecast, is consumed during any period of random length between 1 and 2 days. In the same way, for a VMI customer, the maximum inventory level is chosen as the quantity that is expected to be consumed during a period of random length between 5 and 7 days. An initial inventory level is chosen randomly as an integer number between the safety and the maximum inventory levels. For non-VMI customers, we randomly generated delivery quantities (one for each customer and each product type) in steps of 5 between 5 and 35 tons (i.e., 1 to 7 units). This is justified by a historical, for the studied producer, average delivery size of approximately 20 tons. Hence, for a non-VMI customer the maximum inventory level for a product is assumed to be equal to the safety inventory level plus the delivery size, minus one (deliveries are made so that the inventory level never falls below the minimum level). To be able to compare the results for VMI and non-VMI settings, a maximum inventory level for a VMI customer is not allowed to be less than the maximum level for the same customer and product in the corresponding non-VMI setting. For VMI customers, the inventory levels always have to be kept between the safety and maximum allowed inventory levels, and for a non-VMI customer the inventory level of a product in a particular period equals (i.e., with equality in equation (4)) a certain level that depends on the initial inventory level, delivery size, order point, and forecasted consumption rate.

The time it takes to load a truck is estimated to 1 hour and the time it takes to unload a truck is estimated to 2 hours, with costs 250 SEK and 500 SEK respectively. Before loading takes place at a producer depot, the truck has to be cleaned, and cleaning takes 1 hour and costs 1300 SEK. Loading and unloading times are assumed for full trucks, which we consider to be reasonable since the studied transport operator indicates that it is common that vehicles are fully loaded (or close to being fully loaded).

6.2 VMI Analysis

For the set of 5 randomly generated scenarios, simulation experiments with 3 different VMI settings have been performed. We let *full* refer to a setting where all customers except d_8 use VMI, *none* to a setting with only non-VMI customers, and *one* to a setting where only customer d_3 use VMI. In the experiments we used a termination criteria with parameters $e^V = e^P = 3$ and $g^V = g^P = 10$.

The results from the experiments, with respect to different types of costs, are presented in Table 2. In Table 3 we present the reduction of each type of cost when taking the step from setting *none* to setting *one*, i.e., when going from zero to one VMI customer. The additional cost reductions when letting all customers except d_8 use VMI are presented in Table 4. The economic advantage for *full* compared to *one* is rather

Table 2. The costs for production, transportation, storage, and penalty, as well as the total cost in the system, for each scenario (Sc) and VMI setting. In the penalty cost column, the total penalty cost for violating safety stock levels is written first, followed by the total penalty cost for exceeding maximum inventory levels.

Sc	Setting	Production	Transportation	Storage	Penalty (short / exc)	Total
1	*full*	119000	220741	135738	0 / 0	**475479**
1	*one*	119500	283987	139617	0 / 28720	**571824**
1	*none*	140000	261653	139838	0 / 114800	**656291**
2	*full*	212500	300789	172210	27080 / 0	**712579**
2	*one*	233000	354719	174083	99280 / 65600	**926682**
2	*none*	234500	398659	170829	72220 / 98400	**974608**
3	*full*	166000	262935	174190	0 / 0	**603125**
3	*one*	173500	385060	176325	63140 / 114800	**912825**
3	*none*	145000	366564	163920	198880 / 271480	**1145844**
4	*full*	194500	264527	156875	0 / 0	**615902**
4	*one*	213500	304449	162915	54120 / 4120	**739104**
4	*none*	222500	342993	157340	81180 / 172200	**976213**
5	*full*	134500	252012	160033	0 / 0	**546545**
5	*one*	241000	295026	186117	36080 / 164000	**922223**
5	*none*	221000	305386	178754	63140 / 176320	**944600**

Table 3. The reduction of the production cost, transportation cost, storage cost, total penalty cost, total system cost, and total cost when the penalty cost is disregarded, when going from setting *none* to setting *one*. It should be noted that cost reductions are represented by positive numbers.

Sc	Production	Transportation	Storage	Penalty	Total	Total (no penalty)
1	20500	-22334	221	86080	**84467**	-1613
2	1500	43940	-3254	5740	**47926**	42186
3	-28500	-18496	-12405	292420	**233019**	-59401
4	9000	38544	-5575	195140	**237107**	41969
5	-20000	10360	-7363	39380	**22377**	-17003
Avg	-3500	10403	-5675	123752	**124980**	1228
Std Dev	20338	30917	4711	118332	**102897**	42877

Table 4. The reduction of the production cost, transportation cost, storage cost, total penalty cost, total system cost, and total cost when the penalty cost is disregarded, when going from setting *one* to setting *full*. Again cost reductions are presented with positive numbers.

Sc	Production	Transportation	Storage	Penalty	Total	Total (no penalty)
1	500	63246	3879	28720	**96345**	67625
2	20500	53930	1873	137800	**214103**	76303
3	7500	122125	2135	177940	**309700**	131760
4	19000	39922	6040	58240	**123202**	64962
5	106500	43014	26084	200080	**375678**	175598
Avg	30800	64447	8002	120556	**223806**	103250
Std Dev	43118	33540	10244	74551	**119273**	48755

obvious since it is observed that all costs in each scenario is lower in *full*. However, the economic advantage for *one* compared to *none* is less obvious. Penalty costs represent real costs for obtaining products too early or too late, e.g., by approximating costs for additional storage, wastage, and missed sales. Sometimes, slightly penalized (but cost efficient) plans might be accepted due to the flexibility of the customers. Therefore, it is relevant here to also compare the costs when penalties are disregarded. If penalty costs are included in the comparison of the total costs, *one* always performs better than *none*. However, if the penalty costs are disregarded, *none* sometimes performs better than *one* and sometimes the opposite holds. The average improvement of 1228 SEK does however indicate that *one* works better. In summary, the results indicate that the algorithm manages to find better solutions for the *full* setting than for the *one* and *none* settings.

6.3   Time Performance Analysis

In each iteration, the coordinator agent sends one plan request message to each planner agent (in our scenarios: one transport planner and two production planners), and each planner returns a response message to the coordinator. A response message can either contain a generated plan or a failure notification indicating that no plan could be generated at the time of the request. For our 15 simulation runs, which used an average of 2171 master problem iterations to reach the final solution, in average 13023 messages was be sent. The number of messages can be used as a measure of the overhead in the system in comparison to a non-agent-based implementation approach. The size of each message depends on the actual application and the problem size, since a bigger problem requires that more information (e.g., dual variables) need to be communicated.

For each of the 15 simulation runs, we estimated an upper bound of the overhead imposed by the agentification our decomposition algorithm. This upper bound was calculated as the total running time minus the estimated time for performing decomposition related tasks. We measured an average total running time of 9509 seconds and an average lower bound estimation of 6583 seconds for the time spent in the actual decomposition algorithm. This gives an average estimated overhead of approximately 35.7% of the total running time, which is taken as the average over the estimated overhead of all simulation runs.

Moreover, we estimated the expected performance improvement that can be achieved when all subproblems are distributed and solved in parallel on different computers. For the 15 simulation runs, we measured average estimated total solution times of 3856 seconds for the transportation subproblems and 17.1 seconds for the production scheduling subproblems. Assuming all problems of the same type need the same solution time, we get average times of 0.20 and 0.0039 seconds for solving one transportation and one production scheduling subproblem. The average transportation subproblem solution time is calculated as the average of the average transportation subproblem solution times taken over the 15 simulation runs, and the average production subproblem solution time is calculated in the same way.

If the communication overhead that is caused by a parallelization is disregarded, a theoretical average potential time reduction from $(3856 + 17.1) \approx 3873$ seconds to 429 seconds is estimated for our scenarios. The theoretical time reduction for a scenario is calculated as the number of master problem iterations times the maximum of the average solution time of one transportation subproblem and the average solution time of one production subproblem. In our scenarios, this would give an average time reduction of $3873 - 429 = 3444$ seconds of the total running time. Here, we assume a use of 12 computers; 9 for transportation subproblems, 2 for production scheduling subproblems and 1 for the master problem. Solving the production scheduling subproblems and the master problem on the same computer would give the same improved running time, but confidentiality of information would be weakened. Note that it is impossible to use the main algorithm as it is displayed in Fig. 3 to utilize a complete parallelization. Instead dual variables have to be sent to the transport planners and the production planners in parallel without resolving RMP in between.

7 Confidentiality of Information

An important advantage of the proposed agent-based approach to DW decomposition, compared to a centralized approach, is the possibility to achieve increased confidentiality of information. From the perspective of what type of information need to be communicated, we here provide a comparison between a centralized and an agent-based approach to DW decomposition.

In a centralized approach, a central node of computation needs to be given access to all information that will be used when formulating the problem. From a confidentiality perspective it is not important whether the problem is formulated using decomposition or using a different approach; basically the same information need to be shared. In an agent-based approach, the coordinator needs to obtain all information that need to be used when formulating the master problem, but subproblem specific information can be kept local. The exact information that needs to be shared is case specific, depending on how the problem is formulated with respect to what aspects are modeled in the master problem and what aspects are modeled in the subproblems. Moreover, in each iteration, dual variables need to be sent to the planner agents, who return plans to the coordinator. It follows that complete confidentiality is practically impossible to achieve. For instance, since dual variables and plans are iteratively communicated between the coordinator and the planners, it is often possible for the coordinator to create a model of how planner agents generate plans. However, this has to be done without using explicit information about the planner agents.

In addition to knowing about which depots, production lines and vehicles should be modeled, in our case specific solution approach, the coordinator needs the following information to be able to construct MP (or actually IMP): (1) demand forecast for VMI customers and orders for non-VMI customers, (2) inventory constraints for all depots, and (3) penalty costs (for each depot) for violating inventory constraints. In each iteration, dual variables are sent to the planner agents, who return improving production plans and transportation plans. A production plan (for a particular production line) consists of a price and the amount of each product type that is produced in each period throughout the planning horizon. A transportation plan (for a vehicle) defines a price and the amount of each product type that is delivered to, or picked-up from, each depot in each period during the planning period.

8   Concluding Remarks and Future Work

We have shown that it is possible to create an agent-based approach to optimization based on the principles of Dantzig-Wolfe decomposition. Some rather obvious positive effects of the agent-based approach are increased confidentiality, robustness and possibility of distributed computing. Negative effects are increased processing and communication. However, a parallelization of the approach will typically have a positive effect on solution time and quality. One advantage is the ability to locally specify and modify the subproblems, and another is that a local agent always is able to provide the most recent subproblem solution to the local planners even though it may be based on old dual variables.

In order to capitalize on the use of our multi-agent-based approach for achieving performance improvements, we find it interesting to experiment with a distributed system where agents run on different computers. This is a natural representation of the real world, where actors (agents) are geographically separated and an agent typically only has access to its own data and its own computer. The estimations presented in

Section 6.3 indicate that a parallelization may give a significant improvement on the time performance. Also, a parallelization would allow for solving more complex sub-problems, which would make it possible to capture more details of the actual problem, as well as obtaining solutions with higher quality. Note that our time performance improvement estimation assumes perfect parallelization with no communication overhead, and further investigation needs to be done regarding communication overhead. Regarding the robustness of the approach, planner agents may fail (go down) temporarily, while the coordinator never is allowed to fail. Future work includes refining the algorithmic approach to enable permanent failures of all agent types to be dealt with. For instance, a possible solution to a coordinator failure is to let a planner agent take on the role as coordinator.

In a case study, we have applied the agent-based decomposition approach to a real-world integrated production, inventory and routing problem. The studied problem is rather general and it is independent of the choice of production and transportation sub-problems. However, the planners (subproblems) must be able to take dual variables as input and be able to produce new production or transportation plans that can be communicated to the coordinator (master problem). By choosing customized production and transportation subproblems, our solution approach can be used to solve integrated production, inventory, and routing problems where:

1. Decisions about inventories are taken centrally and detailed decisions about transportation and production are taken locally.
2. The master problem can produce dual variables and receive production and transportation plans.

Hence, we conclude that our approach has the potential to work as a framework for this type of problem. Moreover, the ideas behind our decomposition approach are rather general, and we argue that it can be used to develop solution approaches for other types of problems than our case problem. However, it should be emphasized that a decomposition scheme including master/subproblem formulations, variable restriction strategy, and termination criteria needs to be designed in such a way that the special characteristics of the studied problem is utilized.

For a set of scenarios in the studied case, the presented decomposition approach has been used to conduct a quantitative comparison of different degrees of VMI utilization. The main purpose of the VMI comparison was to illustrate the use of our approach. To obtain results that are statistically significant, more experiments need to be conducted and the approach need to be further validated. The results indicate that the algorithm produces solutions with lower costs for scenarios with more VMI customers. Most likely, the reason is that a higher number of VMI customers increases the solution space of the problem. For instance, we have experienced solutions with high penalty costs for scenarios without VMI customers, which has not been the case for scenarios with VMI customers. A possible explanation is that the algorithm performs better for cases with VMI customers. One reason for this might be that the *full* setting in general uses more iterations than the other settings. Therefore it would be interesting to experiment with termination criteria that allow the different settings to use approximately the same number of iterations before termination. Another possible explanation is the restriction that allows vehicles to wait only in producer depots. This makes it more difficult to obtain transport solutions in which more than one customer is visited in the same outbound trip.

By enabling the economic effects of the introduction of VMI to be studied, we believe that the proposed problem specific solution approach has the potential to provide strategic decision support (concerning VMI) for the involved actors. Moreover, since the approach can be used for suggesting candidate plans for the modeled resources we further believe it has potential to be used as an operational decision support system. For instance, it can suggest plans to human planners in real time to facilitate their decision-making.

References

1. Johnson, E., Whang, S.: E-business and supply chain management: An overview and framework. Production and Operations Management 11(4), 413–423 (2002)
2. Dantzig, G., Wolfe, P.: The decomposition algorithm for linear programs. Econometrica 29, 767–778 (1961)
3. Daugherty, P.J., Myers, M.B., Autry, C.W.: Automatic replenishment programs: an empirical examination. Journal of Business Logistics 20(2), 63–82 (1999)
4. Karageorgos, A., Mehandjiev, N., Weichhart, G., Hämmerle, A.: Agent-based optimisation of logistics and production planning. Engineering Applications of Artificial Intelligence 16(4), 335–348 (2003)
5. Dorer, K., Calisti, M.: An adaptive solution to dynamic transport optimization. In: AAMAS 2005: Proceedings of the Fourth International Joint Conference on Autonomous Agents and Multiagent Systems, pp. 45–51. ACM, New York (2005)
6. Davidsson, P., Persson, J.A., Holmgren, J.: On the Integration of Agent-Based and Mathematical Optimization Techniques. In: Nguyen, N.T., Grzech, A., Howlett, R.J., Jain, L.C. (eds.) KES-AMSTA 2007. LNCS (LNAI), vol. 4496, pp. 1–10. Springer, Heidelberg (2007)
7. Petcu, A.: A class of algorithms for distributed constraint optimization. Ph.D. thesis, Swiss Federal Institute of Technology, Lausanne, Switzerland (2007)
8. Geoffrion, A.M.: Lagrangean relaxation for integer programming. Mathematical Programming Study 2, 82–114 (1974)
9. Benders, J.: Partitioning procedures for solving mixed-variables programming problems. Numerische Mathematik 4, 238–252 (1962)
10. Barnhart, C., Johnson, E.L., Nemhauser, G.L., Savelsbergh, M.W.P., Vance, P.H.: Branch-and-price: Column generation for solving huge integer programs. Operations Research 46, 316–329 (1998)
11. Hirayama, K.: Distributed lagrangean relaxation protocol for the generalized mutual assignment problem. In: AAMAS 2006: Proceedings of the Fifth International Joint Conference on Autonomous Agents and Multiagent Systems, pp. 890–892. ACM, New York (2006)
12. Drexl, A., Kimms, A.: Lot sizing and scheduling - survey and extensions. European Journal of Operational Research 99, 221–235 (1997)
13. Toth, P., Vigo, D. (eds.): The Vehicle Routing Problem. SIAM monographs on discrete mathematics and applications, Philadelphia (2002)
14. Thomas, D.J., Griffin, P.M.: Coordinated supply chain management. European Journal of Operational Research 94, 1–15 (1996)
15. Sarmiento, A.M., Nagi, R.: A review of integrated analysis of production-distribution systems. IIE Transactions 31, 1061–1074 (1999)
16. Federgruen, A., Zipkin, P.: A combined vehicle and inventory routing allocation problem. Operations Research 32(5), 1019–1037 (1984)

17. Chien, T.W., Balakrishnan, A., Wong, R.T.: An integrated inventory allocation and vehicle routing problem. Transportation Science 23(2), 67–76 (1989)
18. Campbell, A., Clarke, L., Kleywegt, A., Savelsbergh, M.: The inventory routing problem. In: Crainic, T.G., Laporte, G. (eds.) Fleet Management and Logistics, pp. 95–113. Kluwer Academic Publishers, Dordrecht (1998)
19. Chandra, P., Fisher, M.L.: Coordination of production and distribution planning. European Journal of Operational Research 72(3), 503–517 (1994)
20. Lei, L., Liu, S., Ruszczynski, A., Park, S.: On the integrated production, inventory, and distribution routing problem. IIE Transactions 38(11), 955–970 (2006)
21. Persson, J.A., Göthe-Lundgren, M.: Shipment planning at oil refineries using column generation and valid inequalities. European Journal of Operational Research 163, 631–652 (2005)
22. Bilgen, B., Ozkarahan, I.: A mixed-integer linear programming model for bulk grain blending and shipping. International Journal of Production Economics 107(2), 555–571 (2007)
23. Fumero, F., Vercellis, C.: Synchronized development of production, inventory, and distribution schedules. Transportation Science 33(3), 330–340 (1999)
24. Davidsson, P., Holmgren, J., Persson, J.A., Ramstedt, L.: Multi agent based simulation of transport chains. In: AAMAS 2008: Proceedings of the 7th International Joint Conference on Autonomous Agents and Multiagent Systems, pp. 1153–1160. International Foundation for Autonomous Agents and Multiagent Systems, Richland (2008)
25. Sohier, E.: Modelling a complex production scheduling problem - optimization techniques. Master's thesis, Blekinge Institute of Technology, Sweden (2006)
26. Bellifemine, F.L., Caire, G., Greenwood, D.: Developing Multi-Agent Systems with JADE. Wiley Series in Agent Technology. John Wiley & Sons, Chichester (2007)

e-Sourcing Clusters in Network Economy

Konrad Fuks and Arkadiusz Kawa

Poznań University of Economics,
al. Niepodległości 10, 61-875 Poznań, Poland
{konrad.fuks,arkadiusz.kawa}@ue.poznan.pl

Abstract. Acquiring resources has always played an important role in doing business. Initially companies gathered them within their own capacity until different sorts of business clusters started to appear. They began to purchase products and services needed to do business by their individual members in order to obtain preferential prices and other trade conditions. Thanks to geographical proximity as well as close and stable cooperation, companies are able to gain a lot of benefits that they would not reach if they worked on their own. The information and communication technologies (ICT) has given a possibility to go beyond traditional cluster concept and has allowed various geographically spread companies to form electronic clusters (e-clusters). Agent technology seems especially promising in this respect. Software agents are able to support inefficient traditional e-marketplaces, including partners who offer one another the best cooperation possibilities and conditions at a given time. The main contributions of this book chapter is the presentation of new resources acquisition method and the e-sourcing cluster concept. Additionally authors have built and verified simulation model of four e-sourcing cluster strategies.

Keywords:, e-sourcing, network economy, clusters, logistics, electronic supply chains, agent based simulation, NetLogo.

1 Introduction

The information and communication technology revolution that has been observed for the last 40 years, and especially the development of the Internet, has initiated a transition of well-developed and developing economies in the direction of an information society. On the basis of this information "order", a new economy called informational, global or network-ized has grown.

The new economy has caused changes in the logic of enterprise organization. Knowledge, unique resources and key competences are being brought to the foreground. Increasing competitiveness is forcing a new organizational model characterized by newer and faster alliances created more frequently. These are formed on an international as well as local scale, between different industries, markets, spheres of activity, or individual enterprises and lead to a network-ized organization model. All functions of an organization which are not part of its basic activity and which may be fulfilled more process and cost-effective by other parties are assigned outside.

A. Håkansson & R. Hartung (Eds.): *Agent & Multi-Agent Syst. in Distrib. Syst.*, SCI 462, pp. 33–51.
DOI: 10.1007/978-3-642-35208-9_2　　　© Springer-Verlag Berlin Heidelberg 2013

The move of the global-economy-functioning axis towards knowledge, unique resources and key competences has caused a selection of new organizational models. They are based on networks and virtuality and constitute a response to the changes brought about by the era of the information society.

Virtuality, with reference to enterprises, is understood as the use of the network character of business connections in reply to the opportunities emerging on the markets. An enterprise in the course of virtualization oversteps its limits in the search of potential partners to enter into cooperation with in order to achieve a specific benefit. These partners do not simply play a traditional role of suppliers but together build a value chain (virtual organization), which creates a product consistent with the needs of the final client.

An electronic cluster, which is used to acquire resources on e-marketplaces, is an example of such new organizational models presented in this article.

This chapter is a continuation of the author's work presented at the KES-AMSTA[1] 2009 and 2010 conferences, regarding resource acquisition by e-sourcing clusters. Its structure is as follows: the evolution of resource acquisition is discussed in Sec. 2, then, Sec. 3 describes the strategic role of sourcing and Sec. 4 presents e-sourcing. In Sec. 5 the idea of clusters is explained and in Sec. 6 the e-sourcing cluster is described. Four e-sourcing cluster strategies proposed by the authors are presented in Sec. 7. Next, Sec. 8 depicts the simulation experiment background, variables, constants and constraints, while Sec. 9 contains the results of the simulation experiments. The conclusions and directions for future research are outlined in Sec. 10.

2   Evolution of Resource Acquisition in Enterprises

Nowadays, the activity of almost every enterprise relies on resources to a large extent; it may even be stated that an enterprise would not be able to function without resources. Resource can be defined as (1) an available source of wealth, (2) a new or reserve supply that can be drawn upon when needed.[1] The above formulation shows that enterprise resources may be very broadly understood. The classification of resources most frequently encountered in literature divides them into: material ones (e.g. buildings, machines, raw materials, products), immaterial ones (e.g. knowledge, information, competence) and human resources (employees). The subject matter of this article concerns material and immaterial resources that may be acquired on electronic markets. Human resources, in turn, are a very specific type of enterprise resources subject to completely different management methods than the other two; therefore, they have been omitted for the sake of cohesion.

Resource acquisition understood as commerce traces back to the beginning of human civilization. The first finds proving commercial exchange date back to Pleistocene. The beginning of commerce may also be related to the emergence of surplus production of the prehistoric human. Awareness of the possibility that the products may be exchanged initiated commerce which was taking the form of barter exchange throughout most of human history.

---

[1] International KES Symposium on Agents and Multi-agent Systems – Technologies and Applications.

In ancient times merchants travelled to distant countries to purchase resources and find new markets. Mesopotamia imported wood, stone, metal ores and precious stones. Egypt provided itself with long wooden beams used in construction. Many cities located at the Mediterranean Sea, on the other hand, needed food and raw materials for craft production. Luxurious goods, such as silk imported from China, were also objects of trade. The Silk Road linking Europe with Asia attracted thousands of merchants avid for the precious fabric and soon became the place where European goods were exchanged for Asian ones. The trade towards the Mediterranean Sea mainly involved silk, paper and iron, whereas gold, cultivated plants and aromatic oils were exported to China.

Sourcing has been subject to numerous alterations since that time. The interest in purchasing started to increase among many scientists in the second half of the 20th century. The influence of resource acquisition on the activity of enterprises began to be investigated thoroughly. In his article, published in Harvard Business Review [2], D.S. Ammer, after having examined 750 enterprises, stressed that high-level management considered purchases to be of passive significance to the enterprise activity. Such an attitude resulted from ignorance and lack of knowledge and understanding what a purchasing process really was and what elements it consisted of. The difference in understanding the process of resource acquisition as well as the lack of directives related to purchasing on the strategic level led to poor effects of the purchasing process. The crisis on the oil market of the 1970s did not change the attitude towards resource acquisition in enterprises, either.[3]

H.I. Ansoff also emphasized that purchases in enterprises (in the 1960s and 1970s) had an administrative, not strategic, character.[4] J.R. Caddick and B.G. Dale noticed that enterprises did not conduct analysis and planning processes correctly, which translated into insufficient strategy preparation and implementation, especially in the sphere of purchase functions.[5]

3   Strategic Attitude towards Resource Acquisition

A closer look at M.E. Porter's work [6] and his five forces influencing competitiveness makes the key role of the purchasing process in enterprise activity, and ipso facto in its strategy, become obvious. It mainly results from two out of Porter's five forces, the bargaining power of suppliers and the bargaining power of customers, that is. Porter considers the choice of an appropriate group of suppliers and customers to be a key strategic decision. Enterprises can improve their strategic position by selecting suppliers and customers characterized by the least force of being influenced negatively.[6] Porter's three remaining forces may also be treated as those exerting influence on the purchasing function of an enterprise. Their influence is not so obvious, yet a strategic approach to resources may greatly contribute to a better competitive position in a given industry (the intensity of competitive rivalry), protection against new entrants (the threat of the entry of new competitors) or a better product / service adjustment to the customers' needs (the threat of substitute products or services). It clearly shows that the concept, so basic for shaping enterprise competitiveness, has become one of the mainstays of the strategic meaning of resource acquisition.

The collective analysis of the research on the strategic role of the purchasing function (see table 1) conducted by L.M. Ellram and A. Carr [7] confirmed the continuously increasing interest of analysts, researchers and enterprises in the problems related to the purchasing function in the enterprise strategy. Growing competition as well as a need to reduce costs and, simultaneously, maintain, or even improve, the quality of products stimulated enterprises to look at resource acquisition in a more strategic way.

Table 1. The role of resource acquisition in supporting an enterprise strategy

Author/Authors	Research methodology	Main conclusions
Spekman	Conceptual	The purchasing function must be included in the enterprise strategy. Resource acquisition must reflect the way of thinking and the strategic development of the company.
Browning et al.	Conceptual	The purchasing function is connected with the enterprise strategy as it supports the monitoring and interpretation of the purchasing trends, the identification of the ways of supporting the company strategy and the development of possible resource acquisition.
Burt and Soukup	Conceptual	If the purchasing function is included in the development of new products soon enough, it may significantly influence the success of their launch.
Caddic and Dale	empirical – case studiem	Resource acquisition must develop strategies and be connected with the overall enterprise strategy.
Landeros and Monczka	empirical – interviews	The purchasing function may support the company's strategic position thanks to the cooperative relations between the buyer and the seller.
Carlson	empirical – case studiem	The purchasing strategy is important from the perspective of product development and the company's long-term objectives.
Reid	Conceptual	Resource acquisition should be included in the early stage of the enterprise strategy development in order to achieve cohesion with the company's strategic plans.
St. John and Young	empirical – survey	Managers responsible for purchases, production and production planning are unanimous as far as the significance of a long-term strategy is concerned. However, their every-day tasks are inconsistent with strategic, long-term plans of the enterprise.

Source: Ellram L. M., Carr A., Strategic purchasing: A history and review of the literature, International Journal of Purchasing and Materials Management, Vol. 30, Issue 2, 1994, pp. 14.

The strategic importance of resource acquisition was also confirmed by European researchers. L. Gadde and H. Hakanson identified three strategic issues related to resource acquisition [8]:

1. An increase in the significance of "buy" over "make". A decrease in vertical integration and greater complexity of the problems related to make-or-buy caused a necessity of a strategic attitude to these issues.
2. Systematic attempts at structuring the supplier base, including their reduction and an improvement of coordination.
3. Closer cooperation with key suppliers in order to improve technological development.

The most important implication of the above issues was a considerably increasing interest of high-level management in the purchasing process and, what follows, growing strategic significance of resource acquisition. Managing relations with key suppliers, designing products together or planning technological investments required strategic decisions from the point of view of the entire enterprise.

Another consequence was purchase decentralization, resulting from two trends in enterprises [8]: 1) decisions should be made by people being closest to the problems in question; 2) creating the so-called independent profit-generating centers which were closely related to the decentralization of the decision-making process. The tendency to outsource was mainly connected to a high level of purchasing costs in the total costs of the enterprise (over 50% [8]). The decision decentralization facilitated entering into closer contacts with suppliers, which entailed better supply conditions and, at the same time, reduced costs of the entire enterprise activity.

Changes in the organizational structures of enterprises took place, too. They were caused by an increase of the strategic aspect of resource acquisition. The dominant selling function was resigned from in favor of functions oriented more towards integrated problem-solving. Thus, logistics and enterprise technological development gained more significance.

The change in the perception of the purchasing function was also reflected on a more global level of enterprise activity. The increased importance of the relations with suppliers resulted in the emergence of a network of connections which expanded the specialization potential of the given enterprise as well as the whole supply chain.

4   e-Sourcing

Strategic sourcing on a global scale, though, requires integration and co-ordination of the supply needs of the enterprise's all economic units from the perspective of the common products, processes, technologies and suppliers.[9] It would not be possible without ICT, especially the Internet, whose dynamic development in the 1990s initiated the idea of e-sourcing.

E-sourcing is defined *"as the use of Web-based applications, decision-support tools, and associated services to identify, evaluate, negotiate, and configure purchases and supplier relationships that will effectively support supply chain and other business operations"*.[10]

At the beginning of the e-sourcing evolution, enterprises built electronic product catalogues, ordered goods via an internet browser and made payments on-line. Now e-sourcing manifests itself in establishing close co-operation between recipients and suppliers as well as in better information and knowledge management.[10]

The benefits of e-sourcing can be divided into operational and strategic ones. The former include purchase costs reduction (discounts resulting from aggregation of purchase, greater competition among suppliers, more possibilities in the field of negotiation and offer analysis, etc.) and process improvement (faster, simpler and more flexible product ordering, continuous accessibility of supply markets, etc.) while the latter are related to increasing the innovative potential of the enterprise (identification of new buying sources and evaluation of the previous ones, incorporating suppliers into innovative processes, etc.).[11]

The accomplishment of the e-sourcing process assumes making use of the transaction-negotiation solutions. These may be: on-line auctions, on-line catalogues, electronic RFx queries and electronic marketplaces (e-markeplaces).

The first of the above methods presumes competition between selected participants who want to conclude a purchase or sale transaction on pre-set conditions. The winner of the auction is the one who has offered the best conditions. The following may be distinguished among on-line auctions: reverse auction, dutch auction, first-price sealed-bid auction, vickrey auction.[12]

On-line catalogues, on the other hand, can be used in two ways: via a website or an electronic purchase system (in this case it is possible for the buyer to combine catalogues of different suppliers, which definitely makes comparing offers easier). Suppliers can establish custom catalogues for buyers who can, in turn, work with pre-established supplier catalogues and prices to procure material and services. There are three different types of electronic catalogue options: static product catalogue, static configurable product catalogue (content is static and has to be updated on regular basis by the vendor), and dynamic product catalogue (content is generated the moment the user accesses the catalogue). When describing on-line catalogues one cannot forget about the most technologically advanced dynamic catalogue which collects data directly from the catalogues on the suppliers' websites.[11]

The third transaction-negotiation solution is electronic RFx queries. These can be divided into: RFI – Request For Information, RFQ – Request For Quotation and RFP – Request For Proposal. RFI is one of the most general ones as it is supposed to initiate receiving information about the supplier and the estimated purchase costs. The next two, RFQ and RFP, are followed by notification about a possibility to make an offer. RFx queries include all requirements related to the supplier, a detailed description of the needed product and the criteria for the choice of the best offer.[13]

The fourth method is e-marketplace which, similarly to on-line catalogues, enable standard goods (usually raw materials) trade. These are internet platforms joining buyers, sellers and brokers in order to increase the competition between suppliers and to obtain access to a greater number of purchasers. E-marketplaces offer dynamic price establishment, aggregation of orders, making offers to numerous suppliers, individual suppliers' price openness, negotiation mechanisms and co-operation possibilities. E-marketplaces are the most sophisticated form of e-sourcing used today and their popularity is still rising. eMarket Services currently has about 640 e-marketplaces in its database.[14] One of the most important advantages of e-marketplaces is that they do help companies find the almost ideal suppliers.[15] E-marketplaces can reduce the potential partner searching costs, as well as the administration costs, on both buyer and seller sides.[16]

Although they offer a lot of facilities and have many strengths, there are some limitations of these solutions supporting e-sourcing. Firstly, today it is very easy to establish a marketplace on the Internet. Thus, there are hundreds of globally operating e-marketplaces. It is very difficult to sort all the information about potential suppliers from so many e-marketplaces and to compare it in order to make a final decision. Moreover, the environment of e-marketplaces is constantly changing – some e-marketplaces occasionally give much better offers than others.[17]

Electronic markets are very important element of modern network-ized economy. Though their adjustment to customers' needs, integration with broad spectrum of other IT systems, or implementation of sophisticated inner negotiation algorithms will among other things determine the future development of e-marektplaces. Some other insights about e-marketplaces in relation to presented in this chapter e-sourcing model are presented in section 6.

5   Cluster

In order to solve the problems presented in the previous section we are suggesting to use the e-sourcing cluster concept to acquire resources needed by companies. Before this concept is presented, however, let us explain what the cluster is, according to the traditional approach.

Generally, a cluster can be described as a group or bunch of objects. Merriam-Webster Online Dictionary defines a cluster as a number of similar things that occur together.[1]

Cluster applications are multiple. The concept may be found in astrophysics (star cluster), biology and health sciences (cancer cluster), computing (computer cluster), music (tone cluster), economics (business cluster), the military (cluster bomb), etc. This paper focuses on the business cluster. Intuitively the business cluster can be specified as a group of companies which cooperate in order to achieve some advantages.

The term, business cluster, also known as an industry cluster, competitive cluster, or Porterian cluster, was introduced and popularized by M.E. Porter in his work, *The Competitive Advantage of Nations*.[18] He defines a cluster as *"...a geographically proximate group of interconnected companies and associated institutions in a particular field, linked by commonalities and complementariness"*.[19]

The best example of a business cluster is The Silicon Valley (in the field of computer technology), Hollywood (U.S. movie production site), and Detroit (U.S. automobile production plants concentration).

Clusters are formed by enterprises linked through vertical (enterprises connected with others in the relation "supplier-buyer") and horizontal (e.g. competitors creating an alliance to achieve the strategic goal) relationships with the main players located in the same place. The geographical proximity is seen as a way of facilitating the transmission of knowledge, supplies and the development of institutions, which in turn may enhance cluster effectiveness.[20] According to M. Porter, it creates competitive advantages for small and medium enterprises (SMEs) which cooperate and compete closely, since a host of linkages among cluster members results in a whole greater than the sum of its parts.[19] So, all the companies can benefit directly from being part of a cluster.

Unfortunately, clusters understood in this way have some limitations. The definition presented above and many other existing in the professional literature and papers [18][20][21][22][23], suggest that the characteristic feature of the business cluster is the geographical proximity. This vicinity as well as the informal communication and face-to-face contacts connected with it still matter and create the competitive advantage.[24] However, not every enterprise (especially SMEs) can be a member of the business cluster due to high entry costs. Moreover, it is impossible to physically locate all enterprises in one place. Further, clusters are formed for years, that is why there is a high exit cost.

6   E-sourcing Cluster

As mentioned above, the problems of resource acquisition on e-marketplaces can be solved by the e-Sourcing Cluster (eSoC) concept. The e-sourcing cluster (a type of a business cluster) is a group of enterprises which are looking for the same type of resource (e.g. steel, plastic, packaging, transportation, etc.), so they cooperate in order to acquire it more quickly and more effectively.

They communicate and exchange information automatically using software agents. In such clusters the geographical proximity does not matter at all. Enterprises do not have to be located in one place because they cooperate using the Internet. The eSoC consists of net of enterprises from the same or similar branch, which are characterized by the capability of quick and dynamic adaptation to the changing market and other requirements. The eSoC concept can be very attractive to SMEs, because it refutes high entry/exit costs barrier from the traditional cluster theory.[25] Another distinctive feature of such clusters is decentralization and the lack of one central point around which enterprises are concentrated.

As e-sourcing clusters have highly distributed, decentralized character, so there is also no centralized information storage. There are many interconnected nodes (software agents) that individually interchange information and form e-sourcing clusters. This idea complies with Digital Business Ecosystems (DBE) paradigm assuming highly network-ized social and business reality. Companies acting within DBE, regardless of their size, have equal access to resources, competencies and knowledge indispensable for their goals and tasks realization. This interconnected, decentralized characteristics of DBE derive from:[26][29]

- resource limitations,
- necessity to avoid having a single point of failure,
- need to more efficient use of distributed information and resources,
- need to increase the performance and stability of the system.

As eSoC model is very closed to DBE or even it can be stated that eSoC can be a part of DBE, so other features of eSoC model corresponds with ones that define DBE. Table 2 shows these characteristics divided into three science domains: social, computer and natural science.

Table 2. DBE characteristics [27]

Social Science	Computer Science	Natural Science
• A community of users • A shared set of languages • A set of regulatory norms and guidelines to foster trust • A population of services • An open-source service-oriented infrastructure	• Several categories of users • A set of formal languages • A security and identity infrastructure • A service-oriented architecture • A service development environment • A distributed P2P run-time environment • A distributed persistent storage layer	• A population of interacting agents/ apps • A distributed evolutionary environment • A dynamic, adaptive, learning, and scale-free network infrastructure

Clustering within DBE can be described as extended, dynamic and knowledge-oriented (type D at figure 1). This perfectly corresponds with eSoC cluster characteristics presented in this section. Within eSoC model all business interconnections can be very dynamic and created even for one transaction. Additionally agents (companies electronic representatives) can participate in multiple virtual clusters. This type of relations in highly virtualized environment can led to global network of cooperation, one correlative economic ecosystem – Digital Business Ecosystem.

Open-source, service oriented character of DBE infrastructure require standardized, interoperable information exchange protocols. When we refer to all types of information exchange (websites, rss, social networks, emails, business messages, etc.) via computer networks, mainly internet, this desired state (combination of protocols) is called semantic web. The access to the semantic web does not require application of any specialized IT systems. Information can be process in format that is readable and understandable simultaneously for computers and humans. This sounds a little bit futuristic but if we take only business information as a subject of exchange we can refer to electronic data exchange standards as desired equivalent for eSoC model information interchange protocols.

Nowadays business information exchange standards originated from combination of EDI (Electronic Data Interchange) standards and XML (eXtensible Markup Language). Combination and extension of above-mentioned led us to multiple information exchange standards: ebXML, RosettaNet, EPCglobal Network, GDSN, BizTalk. Unfortunately there are some interoperability problems between those standards. However it is most likely that one of those standards will occur as leading one and others will have to evolve to interoperate with it or just simply will vanish from e-business reality.

In authors opinion combination of EPCglobal Network and GDSN (both supported by GS1 consortia) are the most promising background for eSoC model information exchange. Main reason for our approach is complementarity of both standards. Global Data Synchronization Network (GDSN) corresponds for static data exchange: localization of company's facilities, product categorization and description, customers basic segmentation, etc. Whereas Electronic Product Code Global Network (EPCglobal Network) stands for dynamic data exchange: serial number of individual product, production date, product status change (shipment, warehousing, placement on store shelf), etc. Proper combination of these two standards enables efficient, unified and transparent information exchange between enterprises.

GDSN is global, internet based, network of independent data pools with central registry managed by GS1. Information within GDSN is stored locally in data pools which are highly distracted over the internet and central registry is only a map of those data pools.

On the contrary EPCglobal Network is responsible for monitoring of dynamic flows in supply chains. It is based on automatic identification technologies (e.g. Radio Frequency Identification) and allows companies to track and trace information about product flows in supply chains. The infrastructure of this Network are EPCISs – Electronic Product Code Information Services – stand alone databases with all company's dynamic data and reference for data pool with static data.

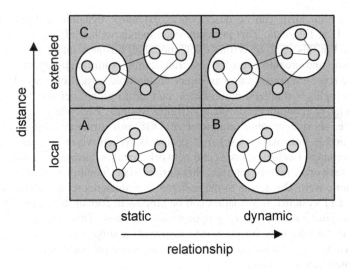

Fig. 1. Clustering typology [30]

This new network-ized business ecosystem can be very promising for SMEs, which can on its basis develop new competitive advantages and new competencies. Within the e-sourcing cluster SMEs can coopete[2] with big enterprises which use their size and purchasing power to get cheaper supplies and offer inexpensive goods. The

---

[2] Coopetition is a neologism which joins cooperation and competition. Examples of coopetition include Apple and Microsoft building closer ties on software development and the cooperation between Peugeot and Toyota on a new city car for Europe in 2005.

entities joined in eSoC share useful information and knowledge with other members in order to achieve better reciprocal understanding, acquire a wider offer range and develop a base for mutual trust that may eventually lead to collaboration in achieving the members' individual as well as collective goals.[25] Thanks to that, it is easiest to reconfigure elements of sourcing processes according to changing output requirements and the rise of a new market.[26]

Obviously, it must be remembered that in the case of eSoC some transaction costs are higher (e.g. the transportation costs are boosted due to the longer distances), but can be compensated by a lower cost of searching and choosing suppliers, contracts management, data interchange, effects of synergy, and, what is most important, lower prices of resources.

Following paragraphs will describe how companies with support of various types of software agents can utilize eSoC model for resource acquisition. Authors assume that companies within eSoC model are familiar with agent and other background technologies that must be utilized to tap eSoC potential.

First of all company must unify its processes according to GS1 standards (GDSN and EPCglobal Network). As eSoC model covers only buy-sell transactions standardization process applies just to small part of company's processes. Secondly company implements on its local server or other personal computer with static IP and internet access eSoC module (extension of EPCIS) with group of software agents, GS1 standards based database for sell and purchases transaction storage, validation mechanisms and other features responsible for data integrity and automatization of e-sourcing processes (Fig. 2).

If configuration of eSoC module and information about company (company's profile) are validated a delegated agents look for e-sourcing cluster/s (eSoC/s). If they do not find one, they create one. In the next step *leader agents* (type of software agents) of each company assigned to the e-sourcing cluster delegate other, subordinate to them, *scout agents* (another type of software agents) to search for resource with defined conditions (quantity, quality, price, shipping terms, etc.) and with proper eSoC strategy (cf. section 7). After that, *scout agents* visit all potential EPCISs (equivalent of individual e-marketplaces) and look for resources (sell offers) that meet the predefined conditions.

If *scout agent* finds proper resource, it compares resource supply (*VRs*) with it company's demand (*Dm*). Than it reserves *Rv* units of the resource, where *Rv* value depends on:

1) $Dm >= VRs => Rv = VRs$
2) $Dm < VRs => Rv = Dm.$

In first scenario only one company acquires the resource. Its demand can by partially ($Dm > VRs$) or fully satisfied ($Dm = VRs$). If $Dm > VRs$ *scout agents* keep looking for next resource offers and its *leader agents* monitor eSoCs' database for findings of other agents.

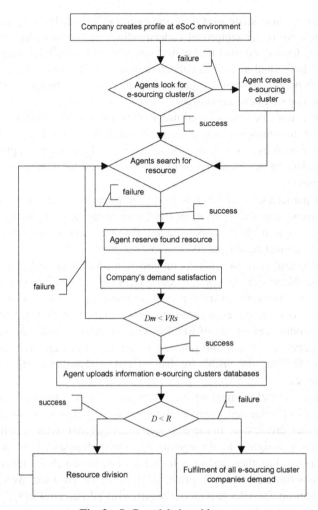

Fig. 2. eSoC model algotrithm

When supply of the resource exceeds demand of *scout agent*, that company satisfies whole demand. Next, *scout agent* comes back to the eSoC from which it originate and uploads information about resource into eSoC database. Than it passes the information to other eSoCs in which it participates. Notifying other agents from all of its clusters about resource results in two main advantages. On the one hand, found resource is almost always used. On the other hand the possibility of resource acquisition increase, if the company is a member of greater number of e-sourcing clusters.[17][28]

In the next step *leader agents* sum individual demands of all, interested in resource, companies in its eSoCs and set cluster demand (D) for found resource. Than D is compared with left resource supply ($R = Vrs - Dm$). Found resources can be divided in two ways. If total demand of the eSoCs for the resource is lower than or equal to left supply of the resource ($D <= R$), all *foraging agents* (third type of software agent presented) can migrate to the specified EPCIS and start the resource acquisition process of needed resource. In this case they fulfill all their demands. In the contrary situation, if $D > R$, authors distinguished two types of resource division methods: *First-come, First-buy* and *Proportionally*.

After that *scout agents* visit again potential EPCIS and look for the resource that meets the predefined condition. They repeat this process as long as their demands are satisfied.

7   e-Sourcing Cluster Strategies

As stated in section 2 and 3 sourcing should be the strategic aspect of modern enterprises. The way of resources acquisition mainly depends on supply/demand relation, resource scarcity, size of suppliers base, technological advancement of the product, resource substitutes, or cooperation level with key suppliers. On the basis of those and other market conditions company defines its sourcing strategy. Last twenty years have shown that sourcing is no longer locally oriented (sourcing from the nearest supplier is time and cost effective) and it should additionally take into account global trends in business management. Development of internet technologies supporting business, outsourcing of processes and competencies are, in authors opinion, the two most important trends that have redefined modern sourcing strategies.

When eSoC strategies ware conceptualized two main assumption ware taken into consideration. First one relates to cooperation between companies. Authors claim that orientation to collective value generation and knowledge acquisition is fundamental for companies competitive advantage and effectiveness increase. This statement is supported by works of Marshall [31], Porter [32], Prahalad [33], Williamson [34], Powell [35] to name just a few. European Commission report [36], inner alia, led authors to form second assumption about eSoC strategies: efficient, supported by information and communication technologies, information flow is crucial in the network economy.

On the basis of this two pillars – information and cooperation – authors have distinguished four eSoC strategies which refer to the level of the scout agent's empathy and the method of resource distribution. Agent's empathy is defined as *willingness to search for resources that meet the requirements of other firms (agents) from the agent's eSoCs*. Each resource has three kinds of fitness: price (*price*), place (*place*) and shipping time (*shipTime*). As eSoCs are established on location basis, the agent's empathy refers to the two extant finesses of the resource [17]: price and shipping time. Thus, the condition that found resource must meet, when the agent is empathic, can be described as:

$$price_i = \max(price_{1,...,j}) \tag{1}$$

$$shipTime_i = \max(shipTime_{1,...,j}) \tag{2}$$

The above equations show that the empathic agent seeks for resources that suit at least one firm from its eSoCs. The $price_i$ and $shipTime_i$ variables are new conditions of the empathic agent and $price_{1,...,j}$ and $shipTime_{1,...,j}$ are conditions of agents concentrated in the empathic agent's eSoCs. The methods of resource distribution that are enclosed in eSoC strategies on one hand correspond to the proportional share with resource acquisition, and on the other hand reflect the tough market law: first-come, first-buy. The four selected eSoC strategies are presented in table 3.

Table 3. eSoC strategy matrix

Resource distribution

		first-come, first-buy	proportional
Empathy	no	*egoistic* (strategy A)	*social* (strategy D)
	yes	*helpful* (strategy B)	*empathic* (strategy C)

8   Simulation Assumptions

Evaluation of each e-sourcing cluster strategy was simulated in the NetLogo environment. NetLogo is a programmable agent based modeling environment for simulating natural and social phenomena. Full description of environment and its capabilities and features can be found at [37]. Worth mentioning is fact that NetLogo works on time basis, so changeability of the simulated model over time can be observed and analyzed. Implicitly time in *Netlogo* is measured in *tick* units. *Tick* can represent any possible real time unit (second, hour, day, year, specific time period, etc.).

As in antecedent experiments [17][28] two kinds of *breeds* were distinguished: resources (*ress*) and companies (*coms*). Thanks to that we could define different behaviors and *agentsets* of both breeds.

The fitness of sought after resource was compared with the fitness of found one according to the rules:

1) *price* of found resource cannot exceed *price* declared by the company or set by empathy mechanism (1).
2) *place* of found resource must be the same as region of company.
3) *shipTime* of found resource cannot exceed *ship-time* declared by the company or set by empathy mechanism (2).

The basic number of potential e-marketplaces was assumed to be 9000. Other parameters, which were constant, are as follows:

- *cnum = 1000* – initial number of companies.
- *rnum = 1000* – initial number of resources.
- *maxResLife = 100* – maximum resource life.
- *maxShipWeeks = 1* – maximum randomization of number of shipping weeks.
- *maxCluNum = 5* – maximum clusters number.

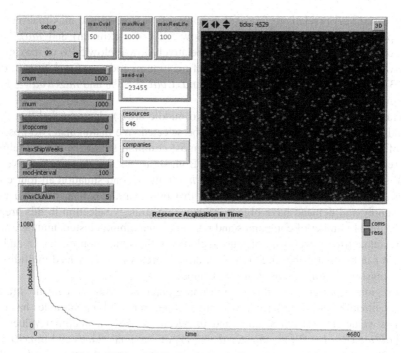

Fig. 3. Netlogo GUI of sample eSoC model simulation

Each strategy was simulated for various values of *maxCval* (maximum individual demand) and *maxRval* (maximum individual supply). Each variable was assumed to take the value between 50 and 1000. Its influence on strategy performance and further on whole population demand satisfaction is the main experiment measurement. Additional important assumptions, for this experiment, describing resources life cycle are:

- Method of new resources generation: every 100 ticks random *rnum* resources are generated – from 0 to 1000 resources.
- Resources are available for random *maxResLife* ticks, evaluated for each resource independently.

The simulation was run 100 times for each case (changing *max-cval* and *max-rval* by 50 and choose one of four eSoC strategies), that gives 160000 simulation experiments.

Analysis of this amount of data needed statistical methods of average values estimation based on distribution of simulation output. Below section refers to average data of eSoC simulation experiment.

9   Results of Simulation Experiments

Analysis of estimated average values showed that the most common is empathic strategy. Being empathic is the best choice in three situations: 1) when maximal individual demand (*maxCval*) is relatively high (above 53% of maximum) and maximal individual supply (*maxRval*) is relatively low (below 53% of maximum) and $maxCval/maxRval \in \langle 0,05; 1 \rangle$. 2) when *maxCval* is relatively low (below 53% of maximum) and *maxRval* is relatively high (above 53% of maximum) and $maxCval/maxRval \in (1; 4)$. 3) both variables are relatively high ($maxCval \geq 53\%$ and $maxRval \geq 53\%$ of maximal values).

Average values visualization (Fig. 3 – two middle areas) also revealed that there are conditions where the best strategy can't be selected. We distinguish two possible states of this situation. At the first area the most common strategies are B (helpful) and C (empathic), but neither is prevailing. Both strategies are empathy positive (cf. table 3) so we called this state *modern empathy*. In the second situation all four eSoC strategies are equally probable. This proves that other variables than maximal individual supply and demand have bigger influence on strategy selection in this area. It might be initial number of companies and resources, maximum clusters number, maximum resource life or even variables not included in the simulation model. In authors opinion it can be interesting direction of future research which may lead multidimensional analysis and verification of the eSoC model.

Last region at presented in this section phase diagram pertain to egoistic strategy (D). This situation can be described with three conditions: 1) both variables has relatively low values ($maxCval \leq 25\%$ and $maxRval \leq 25\%$ of maximal values). 2) $maxCval/maxRval \in (0; 1)$, $maxCval < 25\%$ of maximal values and $maxRval \in (25\%; 30\%)$ of maximal values. 3) $maxCval/maxRval \in (0; 1)$, $maxRval < 25\%$ of maximal values and $maxCval \in (25\%; 30\%)$ of maximal values.

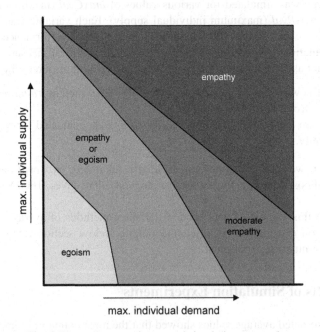

Fig. 4. Phase diagram of averaged simulation output data

10   Conclusions and Future Work

Automatic resource acquisition by enterprises is gaining more and more importance. Companies noticed long ago that cooperation with other parties is better than relying on themselves only. By merging into clusters which base on electronic business and agent technology, companies may get easier and faster access to more attractive sources. The concept of an e-sourcing cluster presented in this paper is a perfect solution for SMEs which, being its member, can respond to the increased pressures of the globalization process and cost reduction, transform themselves and boost their individual competitiveness. Also, they may take advantage of the synergy effects created by entering into cooperative relations with other SMEs and related partner institutions.

The proposed eSoC model can significantly improve the company's resource acquisition process. Additionally, Porterian cluster entry/exit costs lower than traditional ones can indirectly result in a lower risk factor of investing into a new business model. This can, in turn, lead to better cooperation, cross-companies relations and collaboration in achieving the eSoC's members' individual as well as collective goals.

The type of strategy that the cooperating companies choose is, however, of utmost importance. The authors of this paper have suggested four kinds of strategies (i.e. egoistic, helpful, social, and empathic) which depend on the level of agent empathy and the distribution methods. The choice of the best strategy is influenced by the size of resource supply on the market and the size of the demand of the enterprises belonging to the clusters. Certain tendencies have been noticed as a result of the numerous simulation experiments. For example, if the cluster demand is relatively low and the supply – high, the egoistic and social (non-empathic) strategies are more beneficial. Conversely, in the case of high demand and low supply companies benefit most from the helpful and empathic strategies.

Further work on the development of the eSoC model is planned. One of the research directions will be the bundle buying and selling possibilities implementation. Additional work will be conducted in regard to proposing practical insights into the Electronic Product Code Global Network standards as well as the Global Data Synchronization Networks standards.

Another interesting research direction will be to elaborate on the concept of highly decentralized, dispersed individual e-marketplaces. Every enterprise will be able to create such e-marketplaces even for single transactions as well as for longer periods of time.

References

1. http://wordnetweb.princeton.edu/perl/webwn?s=resource
2. Ammer, D.S.: Is your purchasing department a good buy? Harvard Business Review, 36–42, 154–157 (March-April 1974)
3. Farmer, D.: Developing Purchasing Strategies. International Journal of Physical Distribution & Logistics Management 11(2), 114 (1981)
4. Ansoff, H.I.: Corporate Strategy, p. 8. McGraw-Hill Book Co. Inc. (1965)

5. Caddick, J.R., Dale, B.G.: The Determination of Purchasing Objectives and Strategies: Some Key Influences. International Journal of Physical Distribution & Logistics Management 17(3), 15 (1987)
6. Porter, M.E.: On Competition. Harvard Business School Press, Boston (1998)
7. Ellram, L.M., Carr, A.: Strategic purchasing: A history and review of the literature. International Journal of Purchasing and Materials Management 30(2), 10–18 (1994)
8. Gadde, L., Hakansson, H.: The Changing Role of Purchasing: Reconsidering Three Strategic Issues. European Journal of Purchasing and Supply Management 1(1), 33 (1994)
9. Lysons, K., Farrington, B.: Purchasing and supply chain management. Prentice Hall (2006)
10. Aberdeen Group. Making E-sourcing Strategic. From Tactical Technology to Core Business Strategy (2002)
11. Booz Allen & Hamilton. E-sourcing: 21 Century Purchasing (2000)
12. Kaufmann, L., Carter, C.R.: Deciding on the mode of negotiation: To auction or not to auction electronically. Journal of Supply Chain Management 40(2) (2004)
13. Simplexity. Finding outsource resources: The RFI, RFP/RFQ Approach (2003)
14. http://www.emarketservices.com
15. Fuks, K., Kawa, A., Wieczerzycki, W.: Improved e-Sourcing Strategy with Multi-Agent Swarms. In: Conference on Intelligent Agents, Web Technologies and Internet Commerce (IAWTIC 2008). IEEE Computer Society, Washington (2009)
16. Leung, C.S., et al.: Analysis and Design of an Agent Searching Algorithm for e-Marketplaces. In: Cluster Computing (7). Kluwer Academic Publishers (2004)
17. Fuks, K., Kawa, A.: Simulation of Resource Acquisition by e-Sourcing Clusters Using NetLogo Environment. In: Håkansson, A., Nguyen, N.T., Hartung, R.L., Howlett, R.J., Jain, L.C. (eds.) KES-AMSTA 2009. LNCS (LNAI), vol. 5559, pp. 687–696. Springer, Heidelberg (2009)
18. Porter, M.E.: The Competitive Advantage of Nations. The Free Press, New York (1990)
19. Porter, M.E.: Clusters and the new economy of competition. Harvard Business Review 76(6) (1998)
20. Karaev, A., Koh, S.C.L., Szamosi, L.T.: The cluster approach and SME competitiveness: a review. Journal of Manufacturing Technology Management 18(7) (2007)
21. UNIDO, Promoting Enterprise through Networked Regional Development, United Nations International Development Organization, UNIDO Publications, Vienna (2000)
22. Belleflamme, P., Picard, P., Thisse, J.F.: An economic theory of regional clusters. Journal of Urban Economic 48(1) (2000)
23. Pouder, R., St John, C.H.: Hot spots and blind spots: geographical clusters of firms and innovations. The Academy of Management Review 21(4) (1996)
24. Andersson, T., Serger, S.S., Soervik, J., Hansson, W.E.: The Cluster Policies Whitebook. IKED, Malmo (2004)
25. Chung, W.W.C., Yam, A.Y.K., Chan, M.F.S.: Networked enterprise: A new business model for global sourcing. International Journal of Production Economics 87 (2004)
26. Piore, M., Sabel, C.: The Second Industrial Divide. Basic Books, New York (1984)
27. Vidal, M., Gabaldon, J.E.: Digital Ecosystems. Technology and Distributed Nature of Information. In: Nachira, F., et al. (eds.) Digital Business Ecosystems, European Commission, Information Society and Media, p. s.109 (2007),
http://www.digital-ecosystems.org/book/de-book2007.html
28. Fuks, K., Kawa, A.: e-Sourcing Cluster Strategies: Empathy vs. Egoism. In: Jędrzejowicz, P., Nguyen, N.T., Howlet, R.J., Jain, L.C. (eds.) KES-AMSTA 2010, Part II. LNCS, vol. 6071, pp. 312–320. Springer, Heidelberg (2010)

29. Dini, P.: A Scientific Foundations for Digital Ecosystems. In: Nachira, F., et al. (eds.) Digital Business Ecosystems, European Commission, Information Society and Media, p. s.25 (2007), http://www.digital-ecosystems.org/book/de-book2007.html
30. O'Callaghan, R.: Towards Dynamic Clustering: Capabilities and IT Enablers. In: Nachira, F., et al. (eds.) Digital Business Ecosystems, European Commission, Information Society and Media, p. s.71 (2007),
http://www.digital-ecosystems.org/book/de-book2007.html
31. Marshall, A.: Principles of Economics (Revised Edition ed.). Macmillan, London (1920)
32. Porter, M.E.: Clusters and the New Economics of Competition. Harvard Business Review, 77–90 (November-December 1998)
33. Prahalad, C.K., Krishnan, M.S.: The New Age of Innovation. McGraw-Hill (2008)
34. Williamson, O.E.: The Modern Corporation: Origins, Evolution, Attributes. Journal of Economic Literature XIX, 1537–1568 (1981)
35. Powell, W.W.: Neither Market Nor Hierarchy: Network Forms of Organization. Research in Organizational Behavior 12, 295–336 (1990)
36. http://europa.eu/rapid/
pressReleasesAction.do?reference=IP/10/571
37. Wilensky, U.: NetLogo itself. NetLogo. Center for Connected Learning and Computer-Based Modeling. Northwestern University, Evanston,
http://ccl.northwestern.edu/netlogo/

Formalisation and Verification of Knowledge Management in Digital Economy and Organisations

Lilia Georgieva and Imran Zia

Department of Computer Science, Heriot-Watt University, Edinburgh, EH14 4AS

Abstract. Knowledge management is the process of gathering, refining, organising, and disseminating knowledge through which organisations generate value from their intellectual and knowledge-based assets.

In this paper we study standard formal models for knowledge management and discuss how they can be applied to digital economy. We propose a framework for knowledge management applicable to online trading agents. The framework relies on identification of processes which support knowledge management. We express knowledge exchange in epistemic logic. We then translate multi-agent dialogues in a protocol language MAP which is verifiable by model checking.

We apply our framework to a real life case study on knowledge gap and check the underlying knowledge management processes for absence of deadlock and termination.

1 Introduction

In this paper we propose a general framework for verification and simulation of knowledge management processes in an online trading environment. We apply the framework to a real life case study.

We compare our framework to formal models for knowledge management and address the challenge of transferring these models to digital economy.

We define *knowledge management* as the process of critically managing knowledge in order to meet existing needs of a business or organisation, identify and exploit existing and acquired knowledge assets, and develop new opportunities.

In order to develop our framework, we first identify crucial processes which support knowledge management in a digital economy. Next, we formalise the processes in epistemic logic. The formalisation is translated in a multi-agent protocol language *MAP*, which can directly be encoded in *Promela*. The corresponding to *MAP* Promela code is model checked for correctness.

Our contribution is twofold. We first identify fundamental knowledge streams essential for knowledge exchange in digital economy and provide a framework for checking the properties of the exchange. Our framework has the advantage of being provably correct, i.e. we can verify that correctness properties, which correspond to both the behavioural aspects and the data exchange used in the specification of the scenarios of knowledge exchange, are preserved, for example we can check processes for consistency, termination, and fairness.

A. Håkansson & R. Hartung (Eds.): *Agent & Multi-Agent Syst. in Distrib. Syst.*, SCI 462, pp. 53–66.
DOI: 10.1007/978-3-642-35208-9_3   © Springer-Verlag Berlin Heidelberg 2013

The structure of the paper as follows. First, we give background on knowledge management and discuss its importance. Next, we present three of the most popular approaches to modelling knowledge and comment on their advantages and disadvantages in online setting, next we present our framework and finally we demonstrate how our KM framework is applicable to a real life case study.

2   Background

Knowledge is a key asset to any organisation, but organisations are still at an early learning stage of understanding the implications of knowledge management [2]. Consequently, knowledge management is critical for the success of an organisation both in real and digital environment. Economic innovations are frequently direct result of knowledge exchange and sharing. New technical possibilities provide opportunities for efficient operations and yet more effective knowledge sharing.

Digital economy is based on electronic transactions for goods and services produced by an e- business and traded through e-commerce. The transactions are conducted through internet and web technologies and present new challenges in addition to existing ones. Recognised issues in effective knowledge management, associated with significant cost, are for example online security or identity theft. Not less important and costly are issues such as denial of service, or delay of service online. In this context effective and secure knowledge management is paramount.

In this paper we analyse the underlying knowledge management processes and discuss an approach to proving their correctness. We clarify that we take a process-based, rather than project-based perspective to knowledge management. In the process-based perspective, an organisation measures the value of knowledge assets and the impact of knowledge management and then accesses the knowledge from external sources in order to generate new knowledge. Consistent with this approach, we view knowledge management as a collection of suitably identified processes.

Standard knowledge management models utilise the concepts *tacit* and *explicit* knowledge. Tacit knowledge cannot be codified, but can only be transmitted via training or gained through personal experience. It involves learning and skill but not in a way that can be written down.

Explicit knowledge is codified knowledge that can be transmitted in formal, systematic language. It is captured in records and is assessed on a sequential basis. It can be expressed in words and numbers and shared in the form of data, scientific formulate, and specifications, manuals. This kind of knowledge can be readily transmitted between individuals formally and systematically [10].

Formal models, which allow for non-digital knowledge management, have been proposed. We next discuss three prominent models. They are the spiral model [8], the VDT model [12], and computational model [5]. The models are developed knowledge flow in non-digital economy. We discuss the challenges and issues are faced when attempting to model knowledge management in digital economy.

2.1 Spiral Model

The spiral model describes dynamic interaction between tacit and explicit knowledge along an epistemological dimension and characterises four processes of knowledge management, i.e. socialisation, combination, externalisation and integration. The processes enable the individual knowledge to be amplified and thus affect the organisational knowledge [8].

The characterisation of knowledge flow dynamics remains static in terms of this representational model; the model does not use a dynamic representation of knowledge flow, and consequently important dynamic interactions between model elements remain obscured through descriptive models based upon natural language texts and figures [8].

The framework of digital economy delivers a different purpose. Dynamic evaluation of processes is crucial for online transactions. For example, in an online trading environment, supported by eBay or Amazon market place, the winning bid is continuously re-evaluated and changed depending on the time limit, bided amount or even a number of participating agents.

2.2 VDT Model

The VDT modelling environment relies on information processing and embedding it in software tools [12]. Similarly to the multi-agent approach in the VDT environment knowledge exchange partners are modelled as actors communicating with one another and performing tasks.

The qualitative behaviours of VDT models correspond closely to an enterprise processes when put in practise. This is done by embedding software tools that can be used to design organisations.

Multiple virtual prototypes of the knowledge management system are modelled and analysed. The VDT model emulates the dynamics of the organised behaviours.

This approach is transferrable to online environment. A challenge is the flexibility in an organization in which actors can communicate with anyone they choose inside or outside their local "organization". As a consequence of such flexibility many prototypes can be created and the complexity of the validation of such model can be significantly increased. Such organizational form will have links which form and either persist or dissolve in cyberspace.

The validation in a VDT model is achieved via a representation of knowledge processes through simulation which gets increasingly more complex as more agents and behaviours are involved.

2.3 Computational Models

Conventionally computational models describe properties of the knowledge flow independently of the environment [5]. Knowledge flow is associated with complex processes; for example, within software development knowledge flow can be simplified as process types with specific properties.

Computational models are often based on modal or epistemic logics [5]. For reasoning purposes and to ensure reasonable complexity or decidability, the agents are often resource bound and have restricted logical capabilities, e.g. time, memory. Feasible reasoning is ensured by weakening the epistemic logic, i.e. the agent only knows some "obvious" logical truths but not the complicated ones. From this we can assume that the agent can draw all "obvious" consequences, but not any arbitrary consequence, i.e. the deduction mechanism of the agents is not complete. The agents are thus not powerful enough to draw all the logical consequences of their knowledge, which can be challenging in digital economy.

Nevertheless, for simplified trading behaviours, epistemic and modal logic can capture knowledge exchange and is a suitable framework for modelling of agents' behaviour.

3 Framework for Knowledge Management

In this section we present a process-based framework for knowledge management. In the framework we can model capturing and sharing of knowledge among agents.

The framework consists of structured streams which correspond to processes integral to knowledge exchange and can be valided using model checking. Model checking is fully automatic verification technique.

In the context of knowledge management model checking ensures correctness and termination of the underlying processes. The added benefits in a model checked knowledge management framework is that non-terminating processes and processes with deadlocks will be detected and counterexamples generated.

Consider for example a scenario where agents bid for goods online [14]). The protocol assumes a single auctioneer agent and a variable number of bidder agents, which is a realistic scenario for online trading agents (for example on Ebay). The auction begins with the auctioneer sending out the starting value for a particular auction item. Each bidder then makes an internal decision whether to bid at the current value, and makes a bid if appropriate. When the auctioneer receives a valid bid, if there is a higher bidder, the bid value is incremented and the new value is sent to all of the bidders. The bidders then make a decision to bid at the new value. The auction continues until no further bids are received by the auctioneer and a timeout occurs. At this point the winning bidder is notified and the auction concludes.

We use epistemic logic to formalise the exchange of information and bids among agents. We simulate a real-life online trading and identify whether a delay or a deadlock in the trading among agents occurs using SPIN [6].

We identify the following processes, which our framework comprises: determining knowledge and available knowledge, determining knowledge gap, knowledge lock, knowledge sharing, knowledge utilisation and evaluation.

We next briefly discuss each of the knowledge processes that we can model.

Determine Knowledge. Determining knowledge is important prior to any knowledge exchange as it includes the knowledge necessary for the exchange to happen. Consistency checks with respect to the objectives of the exchange to be determined are also part of this process.

Issues to be addressed when transferring determining knowledge to digital economy are: the kind of knowledge compulsory for the exchange (for example unique id or credit card guarantee) and the security constraints imposed on it; the type of knowledge that is needed in order for the exchange to happen (for example password, PIN or other type of identification); the goals of the exchange (for example transfer of goods or services).

Determine the Available Knowledge. This process looks at what knowledge is already available and its strategic implications of, for example, the direction of the exchange in online trading environment. In digital economy, this translates to, for example, keeping track of goods purchased by customers and basing future recommendations on previous purchases. More advanced software allows for personalised recommendations and statistics (currently in use for example by Amazon), where recommendations for future purchases or rentals are based on purchase history.

Determine the Knowledge Gap. The space between the current knowledge and target knowledge defines knowledge gap. The objective for successful business development is that current knowledge equals target knowledge. There are two ways this can happen: the current knowledge can be increased or the target knowledge reduced. Online applications, handling customer details aspire to both. Increasing current knowledge is achieved via cookies that store information of the transactions made by a customer and building a reliable customer profile. Reducing the knowledge amounts to targeting the customer with specific to his interest goods or services and hence reducing by selection.

Knowledge Lock. Knowledge lock in an enterprise means that the business have retained the knowledge into position, e.g. type of software or a type of system in place. Locking the knowledge in digital economy is achieved by storing user profiles unchanged until a new transaction or search takes place, hence updating the user details and preferences.

Knowledge Sharing. Standardly knowledge is a resource for preserving valuable heritage, but also for learning new ideas, solving problems, creating core competences, and initiating new situations for both individual and organisations now and in the future.

Knowledge sharing means distribution of knowledge in the organisation which includes locked knowledge. It is vital that correct knowledge gets to the right person at the right time. This is dependent on the culture and structure of the organisations, arguably distribution of knowledge is straightforward in organisations with flat structure such as Microsoft or Google. In digital economy knowledge sharing can be achieved effortlessly, provided that the correct security checks are in place. A recognised challenge is secure communication and information exchange.

Knowledge sharing in digital economy amounts to knowledge exchange and creation and enhances the competitive advantage of an organisation.

Knowledge sharing in organisations depends not only on technological means, but is also related to behavioural factors. Cognitive and motivational factors are thought to interfere with knowledge sharing both online and in standard face to face communication.

Knowledge Utilisation. Utilisation of knowledge is largely based on the culture of the organisation. Crucial for online economy and trading is the speed and security of the exchange.

Knowledge Evaluation. The evaluation of knowledge concerns knowledge which is already retained in the organisation, and can be utilised. Knowledge can be evaluated e.g. through auditing, conducting satisfaction studies and benchmarking various parts of services [9]. Such evaluation is continuously carried out online.

3.1 Epistemic Logic

In this section we give a brief introduction to our modelling language epistemic logic, based on modal logic.

Epistemic logic is core to the development of agent communication languages [4]. The logic has sound formal semantics and high expressive power, it can be enriched for example by operators modelling commitment, intention, or trust which are integral to business communication in real or digital world.

The semantics of epistemic logic is based on modal logic semantics. Indexed knowledge modalities capture the properties of knowing by an individual agent, i.e. the fact that the agent i knows a fact ϕ is modelled by the epistemic formula $K_i\phi$.

Informally a modal epistemic logic for N agents is obtained by joining together N modal logics, one for each agent. For simplicity it is usually assumed that the agents are homogeneous, i.e. they can be described by the same logic. So an epistemic logic for N agents consist of N copies of a certain modal logic [5]. The following axioms and rules of inference are applicable to modal epistemic logic.

- (PC) All propositional tautologies
- (K) $K_i\alpha \wedge K_i(\alpha \rightarrow \beta) \rightarrow K_i\beta$
- (MP) Modus ponens: from α and $\alpha \rightarrow \beta$ infer β
- (NEC): From α infer $K_i\alpha$

Extensions of the logic can express properties such as the truth axiom (T) $K_i\alpha \rightarrow \alpha$ which states that knowledge must be true or consistency of knowledge (D) $K_i\alpha \rightarrow \neg K_i \neg \alpha$ requiring that agents are consistent in their knowledge, awareness of knowledge or lack of it (4) $K_i\alpha \rightarrow K_i K_i\alpha$; (5) $\neg K_i\alpha \rightarrow K_i \neg K_i\alpha$.

We model multi-agent dialogue in epistemic logic and we translate the model to MAP. MAP is a lightweight dialogue protocol and is used to represent multi-agent dialogue. MAP a replacement for the state-chart representation of protocols in Electronic Institutions and is designed by Foundation for Intelligent

Physical Standardisation (FIPA 02) and is suitable for online information exchange. It permits only the agents that are involved in the scene to take place and excludes the ones that are not relevant.

Conveniently for digital economy additional security measures may be placed or introduced into the scene for e.g. placing entry and exit conditions on the agent. If we place a barrier condition on the agents, a knowledge exchange cannot begin until all the agents are present and the agents cannot leave the scene until the dialogue is complete. The protocols in the knowledge exchange are constructed from operations known as actions.

This controls the flow of the protocol and this also controls the actions which have side effects and can result in failure. Interaction is performed between agents by exchanging messages [13].

The semantics of message passing corresponds to reliable, buffered, non-blocking communication. Sending a message or sharing a message will succeed immediately if an agent matches the definition, and the message (share) will be stored in a buffer on the recipient. Receiving a message requires an additional unification step. The message supplied in the definition is treated as a template to be matched against any message in the buffer.

The foundation of intelligent protocol agents (FIPA) semantics sets the standard for the MAP protocol. FIPA agents communication language (FIPA ACL) is one of the most popular agent communication languages. In this language the interaction between agents is based on the exchange of messages. This defines the sets of performatives known as message types that express the intended meaning of the message. The language does not define the actual content of the message but assumes a reliable method of message exchange.

FIPA ACL message contains a set of one or more message parameters. Precisely which parameters are needed for effective agent communication varies according to the situation; the only parameter that is mandatory in all ACL messages is the performative, although it is expected that most ACL messages will also contain sender, receiver and content parameters.

In order to prove that the knowledge exchange among agents is reliable, and non-blocking we encode MAP in Promela using the approach developed in [13].

The translation into MAP is not new. Our contribution is in using the MAP encoding of formally defined epistemic processes in knowledge management setting. We benefit from using a language that has well-defined syntax and semantics and is especially designed to model properties of knowledge exchange and agent communication. Alternative specifications can be used. For example, in [7] a specially designed specification is used for verification of business processes.

We then model-check the translation. There are a number of similarities between MAP and Promela, for example, both are based on the notion of asynchronous sequential processes or in our case (agents) both assume that communication is performed through message passing or exchange of information to the high-level similarities significantly simplify the translation, as the translation is from MAP agents directly into Promela processes and agent communication into message passing over buffered channels. However, the translation of the low level details of MAP is not straight forward. The difference is that there are significant semantic differences in the execution behaviour of the languages, for example the order of execution of the statements. MAP assumes that messages can be retrieved in an arbitrary order (by unification) while the Promela enforces a strict queue of messages. Using model checking, we can ensure, for example, that the knowledge sharing process is terminating when modelling online trading agents simulating behaviour on online market places.

4 Case Study: Knowledge Gap

We will now apply our framework to a real life case study of knowledge gap, we will first give a background of the knowledge gap case study and demonstrate how our framework is applicable. We will go through each step of the framework and explain the each stage of our KM framework.

We provide a brief background to the knowledge gap case study, which focuses on problems occurring in the Data Networking group at Transport Trails Incorporated (TT Inc.). The case study covers the following areas: (1) definition of the problem, (2) an analysis of the problem, (3) conclusions that propose and discuss potential solutions to the problem.

The company looked at what it knew and what knowledge was available. The main issue is the lack of or no training given to employees [1].

The results from this case study [1] showed that focusing on perceived control, employee loyalty and employee involvement, employees appeared to be satisfied with their jobs. Of those surveyed, 76% felt that they had a lot of control over their jobs, 76.5% felt that they were valued by the organisation and 72% felt that they had a voice with respect to job involvement. The majority of those surveyed felt that they were given the opportunity to influence decisions, systems and procedures, and had the authority to correct problems when they occur without being reprimanded for making mistakes.

Employee involvement encompasses such popular ideas as employee participation or participative management, workplace democracy, empowerment and employee ownership [1]. It can be defined as a participative process that uses the entire capacity of employees and is designed to encourage increased commitment to the organisation's success. The underlying logic is that by involving workers in those decisions that affect them and by increasing their autonomy and control over their work lives, employees will become more motivated, committed, productive and more satisfied with their employee involvement, employee loyalty, all seem to correlate positively with respect to productivity and job satisfaction [3].

The survey conducted in the knowledge gap case study in the IT department [1] found that 70.5% of the employees felt that they were not adequately trained in dealing with the wide range of technical problems that they are likely to encounter. From the open-ended questions received, this variable appears to be one of the biggest obstacles hindering job satisfaction and productivity. Among the comments received were: No time for training although it has been budgeted. Not enough training time to understand new technology. Building up the team, delegating work, therefore time to train is a limited resource.

Essentially, employees feel that there is not enough training and that management needs to evaluate the training requirements of the IT department and ensure that employees acquire the requisite technical skills to perform their duties effectively. What is even more disturbing is the fact that there is no time for training even though it is budgeted for.

Each employee in the IT department is allocated a fixed amount for training and it is left at his/her discretion to ascertain which course(s) he/she requires to improve his/her technical skills [11].

Management needs to map out clearly the training needs for these employees and see to it that their skills are upgraded. Moreover, if time is allocated for training, this may help employees feel more valued. Productivity most probably would increase as well, given that employees will now possess better internal communication skills [1].

Transport Trails Incorporated (TT Inc.) has various issues in its internal communications within its IT engineering department, resulting in IT engineers leaving the company due to low moral and esteem, and overall dissatisfaction with the company. Most of the employees who left, created a major knowledge gap within the company. An external consultancy agency was brought in to identify and rectify the company's problems and to suggest a strategy to rectify them. In the case study [1], a list of alternatives and recommendations to meet the objectives was proposed by the consultancy firm.

4.1 Definition of the Problem

In 2001, one of the two network architects in the Data Networking group decided to leave TT Inc. for another company. Within a month, the remaining network architect also left. Since these individuals had specialised knowledge and skills that could not be replaced internally, their departures created a major critical knowledge gap. The remaining team members in the Data Networking group also shared their dissatisfaction.

Management was faced not only with the task of filling the open positions and closing the knowledge gap, but also addressing the downward-spiralling effects of the situation impacting on the whole information technology department, due to the staff shortage. The workload for the remaining team members increased as TT Inc. attempted to overcompensate for the loss; but at a certain point, the bubble burst. All ongoing IT projects of the Data Networks group ground to a halt. At this point, team members admitted they were overworked, underpaid, unappreciated and that management was not listening. Also contributing to the situation was the news that two key individuals from the closely aligned Telephony Networking group suffered from burnout and were taking extended sick leave.

Productivity and job satisfaction at TT Inc. have declined due to the following: excessive workload coupled with unrealistic deadlines, too much bureaucracy, and a lack of management commitment. The purpose of the survey was to identify the factors having an impact with respect to the organisational problems at TT Inc. The sample consisted of one manager, five team leaders and

11 employees from the Data Networking group. The average number of years of professional experience for the entire sample was 13.5. More importantly, the average number of years worked at TT Inc. for the sample as a whole was 8.5, with a sizable range of 28.5 years stemming from a maximum of 29 years at TT Inc. to a minimum 0.5 years. The next section will discuss the stepwise modelling for the knowledge gap case study.

4.2   Stepwise Modelling

In our proposed framework for knowledge management, knowledge exchange is expressed in epistemic logic, and multi-agent dialogues are translated into a protocol language, which is verifiable by model-checking. The knowledge gap within the agent's knowledge is simulated and the process checks for inconsistencies, termination and fairness.

4.3   Step 1 – Selection of Streams

Step one is where the company selects instruments from the questionnaire, allowing determination of which streams need to be formalised for the internal activities. In this case study, the instruments are already known: brainstorming, developing scenarios, discussion with customers/clients and knowing current market fact sheets, and all of these are knowledge gap instruments.

4.4   Step 2 – Formalisation Process

Once the company has selected the instruments and the stream has been identified, the next step is to formalise the knowledge gap stream. This is known as the formalisation stage. For information on the language used for formalisation.

The knowledge gap stream will now be formalised by using the knowledge law axioms. The first section describes how many agents will be involved in the process and what facts are already known or are available:

- Agent i is the manager
- Agent j is the employee
- Fact α = Current knowledge training plan
- β = Fact (New knowledge training plan)

Formalisation of Knowledge Gap. The formalisation starts with the facts the agents know, that is, their initial states, then go through the formalisation process. This will result in one of the agents knowing what they did not know before, receiving that information from the other agent.

Initial state

- $K_i\ \alpha$ – Agent i knows fact α.
- $K_j\ \beta$ – Agent j knows fact β.

Process Example of processes (intermediate states in the model)

- $K_i \, \alpha \rightarrow K_i \, K_i \, \alpha$.
  Agent i knows α, which implies he knows that he knows α.
- $K_j \, \beta \rightarrow K_j \, K_j \, \beta$. Agent j knows β, which implies he knows that he knows β.
- $\neg \, K_i \, \beta \rightarrow K_i \, \neg \, \beta$. It is not the case agent i knows β, which implies agent i knows not β.
- $\neg \, K_j \, \alpha \rightarrow K_j \, \neg \, \alpha$. It is not the case agent j knows α, which implies agent i knows not α.
- $K_i \, \neg \, \neg \, \alpha \rightarrow K_i \, \alpha$. Agent i knows that it is not the case not α, which implies agent i knows α.
- $\neg \, K_i \, \neg \, \alpha \rightarrow K_i \, \alpha$. It is not the case agent i knows not α, which implies i knows α.
- $\neg \, K_i \, \neg \, \alpha \rightarrow K_i \, \alpha$. It is not the case agent j knows not β, which implies agent j knows α.
- $K_i \, \neg \, K_j \, \alpha \rightarrow K_i \, K_i \, \neg \, K_j \, \alpha$. Agent i does not know agent j knows α, which implies agent i knows that he knows agent j does not know α.

End state

- $K_i \, K_j \, \alpha \rightarrow K_j \, \alpha$. Agent i knows that agent j knows α, which implies agent j knows α.

4.5   Step 3 – Conversion to MAP Encoding Knowledge Gap

The MAP encoding is shown in Figures 1, which give a graphical representation of the process. After the formalisation stage for the knowledge gap, it is converted to MAP encoding then to PROMELA for model checking. The protocol in Figure 1 shows a knowledge gap protocol description between two agents, the manager and the employee, showing the gap being fulfilled.

```
1.    knowledge Gap [
2.    % employee
3.    method() =
4.    wait for
5.    (offer(training) < = agent ($manager %manager) then
6.    call (deliberate, $training, %manager))
7.    timeout(e)
8.
9.    method (wait, training)=
10.   wait for
11.            accept(training) < = agent(employee, %employee)
12.         or reject(training) < = agent(employee, %employee)
13.
```

Fig. 1. Knowledge gap protocol in MAP

The knowledge gap protocol is defined using MAP syntax, as in Figure 1. There are two roles: the %manager and the %employee. The employee (agent) has the option to accept the training or to reject it. Each of the roles has associated methods, which define the protocol states for the roles.

The knowledge gap begins with an offer from the manager to the employee, which is denoted with the message offer (manager, employee). Upon receipt of the initial offer the employee enters a state in which a decision is required, the offer can be accepted or rejected, in which case the protocol terminates. The knowledge gap is effectively captured by a sequence of proposals between the manager and employee.

The MAP semantics for FIPA provides a formal semantics, for the inform performatives expressed in communication in the protocol. It states the proposition is the content of the message, and the sender would like the receiver to accept that proposition marked as α. Whether or not the receiver does accept the proposition, α, will be a function of the receiver's trust in the sincerity and reliability of the sender, from the receivers viewpoint, receiving an inform message entitles it to know that, the sender knows the proposition that is the content of the message, and, the sender desires the receiver to know that proposition also. The end result states the expected effect by the sender after the act is performed, indicated by RE in Figure 2. This states agent i (the sender) will inform j (the receiver) of a proposition or a "content". The term FP is the feasibility precondition and states that if agent i knows a proposition, he knows that he knows the proposition and agent i knows agent j does not know the proposition. RE is the Rational Effect, meaning the effect expected by the sender after the act has been performed and what the sender of the message hopes to achieve. In this case RE states agent i has successfully transmitted the message to the receiver agent j. No reply is needed from the agent j as this is a one-way process.

< i, inform (j, α) >
FP: $K_i\ \alpha \wedge K_i\ K_i\ \alpha \wedge K_i\ (\neg\ K_i\ \alpha) \wedge U \neg\ K_i\ \alpha)$

RE: $K_j\ \alpha$

Fig. 2. FIPA semantics

The next section discusses step 4, the Promela language used for model-checking in SPIN.

4.6   Step 4: Translation of MAP to Promela

An explanation of the Promela encoding has been given, matched against the line number of the code, which is the model-checking process where SPIN checks for successful termination, safety and fairness.

Firstly, the code in Figure 3 is translated. The process shows one-way communication between two processes, the **manager** and the **employee**. The **manager** process sends the value 2 in its local channel variable to the employee channel via the global channel, and so makes it available to that process. The employee process transmits a message of the proper type via a rendezvous handshake on that channel and both processes can be terminated. When the employee process dies, channel loc is destroyed and any further attempts to use it will cause an error.

Line 1 of Figure 3 declares a symbolic name, msgtype of mtype. Line 2 declares a channel called glob which is global to any process and can store one message. In line 3 a process is created. The prefix active means that the process is active. In line 4, a local variable is declared and initialised to 0. Line 5 sends it to the other channel, that is the local channel of line 6, which receives the message. In line 7 a second process is created. The prefix active indicates that a process of type employee is instantiated. Line 8 declares channel who. Line 9 retrieves a message from channel who. Line 10 receives a message value type.

```
1.  mtype = { msgtype }
2.  chan glob = [2] of { chan };
3.  active proctype manager()
4.  {chan loc = [1] of { mtype, byte };
5.  glob!loc;
6.  loc?msgtype(121)
    }
7.  active proctype employee()
8.  { chan who;
9.  glob?who;
10. who!msgtype(121)
11. }
```

Fig. 3. Knowledge gap process in promela

A process of knowledge-gap generated in SPIN has 3 states. The direction of the arrows between states shows the flow of information between the Manager and Employee. The red rectangle indicates the initial stage of the process. In state one, the manager sends a message to the employee, indicated by (glob!loc) . The symbol ! indicates that the message is being sent. The second state shows that the employee is receiving the message, indicated by (loc?msgtype) . The symbol ? means that the message is being received. The final state is 0, which means that the message has been successfully received and the process has terminated.

5 Conclusion and Future Work

In this paper we presented a formal framework for modelling and verification of knowledge management. The framework allows for modelling of properties of knowledge management processes. The processes are checked for consistency, termination and fairness using the model checker SPIN.

The framework can be applied to a range of online knowledge management processes.

We carried out experiments with SPIN for knowledge management dialogues.

In future work we plan to model and model check processes relevant to common knowledge.

References

1. Appelbaum, S., Shapiro, B., Danakas, H., Gualtieri, G., Li, L., Loo, D., Renaud, P., Zampieri, N.: Internal communication issues in an it engineering department. Corporate Communications an International Journal 09(01), 6–24 (2004)
2. Beijerse, R.: Knowledge management in small medium sized companies: Knowledge management for entrepreneur. Journal of Knowledge Management 4(2), 162–179 (2000)
3. Cacioppe, R.: Structured Empowerment: an award winning program at the burswood resort hotel. Leadership and Organisation Development Journal 05(05), 233–254 (1998)
4. Cohen, P.R., Levesque, H.J.: Intention is choice with commitment. Artificial Intelligence 42(2), 213–261 (2000)
5. Halpern, J., Moses, Y.: A guide to completeness and complexity for modal logics of knowledge and belief. Journal of Artificial Inteligence 54, 319–380 (1992)
6. Holzmann, J.: The SPIN Model Checker. Transaction on Software Engineering 23(5) (May 1997)
7. Janssen, W., Mateescu, R.: Verifying business processes using spin, 1998. In: Proceedings of the 4th International SPIN Workshop (1998)
8. Levitt, R., Nissen, M.: Dynamic models of knowledge flow dynamics. Centre or Integrated Facility Engineering 76 (2002)
9. Nonaka, I., Teece, D.: A unified model of dynamic knowledge creation. Journal of Managing Industrial Knowledge 33(1), 5–34 (2001)
10. Polanyi, M.: Knowing and Being. University of Chicago Press (1969)
11. Robbins, P.: Organisational behaviour, p. 481. Prentice-Hall, Englewood Cliffs (2001)
12. Stanford University. The virtual design team research group
13. Walton, C.D.: Model Checking Agent Dialogues. In: Leite, J., Omicini, A., Torroni, P., Yolum, p. (eds.) DALT 2004. LNCS (LNAI), vol. 3476, pp. 132–147. Springer, Heidelberg (2005)
14. Walton, C.: Model checking for multi-agent sytems. Journal of Applied Logic, Special Issue on Logic-Based Agent Verification, 1–28 (2006)

Agent-Based Artificial Immune Systems (ABAIS) for Intrusion Detections: Inspiration from Danger Theory

Chung-Ming Ou[1], C.R. Ou[2], and Yao-Tien Wang[3]

[1] Department of Information Management, Kainan University, Luchu 338, Taiwan
cou077@mail.knu.edu.tw
[2] Department of Electrical Engineering, Hsiuping Institute of Technology,
Taichung 412, Taiwan
crou@mail.hit.edu.tw
[3] Department of Computer Science and Information Engineering,
Hungkuang University, Taichung 433, Taiwan
ytwang@sunrise.hk.edu.tw

Abstract. Agent-based artificial immune system (ABAIS) is applied to intrusion detection systems(IDS). The intelligence behind ABIDS is based on the functionality of dendritic cells in human immune systems. Antigens are profiles of system calls while corresponding behaviors are regarded as signals. ABAIS is based on the danger theory while dendritic cells agents (DC agent) are emulated for innate immune subsystem and T-cell agents (TC agent) are for adaptive immune subsystem. This ABIDS is based on the dual detections of DC agent for signals and TC agent for antigen, where each agent coordinates with other to calculate danger value (DV). According to DVs, immune response for malicious behaviors is activated by either computer host or Security Operating Center (SOC).

1 Introduction

The Internet has become a major environment for propagating malicious codes, in particular, through web applications. Internet worms spread through computer networks by searching, attacking and infecting remote computers automatically. The very first worm is the Morris worm in 1988; while Code Red, Nimda and other Internet worms have caused tremendous loss for the computer industry every since. Recently, OWASP issued the Top 10 vulnerability list for Web applications [1]; intrusions, in particular, web-based ones, have become increasing threats for information assets. In order to defend against network attacks, one efficient way is to understand various properties of virus, which include the impact of patching, awareness of other human countermeasures and the impact of network traffic, even the ways how these malicious codes reside in a certain hosts, etc.

A. Håkansson & R. Hartung (Eds.): Agent & Multi-Agent Syst. in Distrib. Syst., SCI 462, pp. 67–94.
DOI: 10.1007/978-3-642-35208-9_4 © Springer-Verlag Berlin Heidelberg 2013

1.1 Related Work

Forrest et al. [2] proposed an instance of computer immunology to protect the computer systems using the principles of human immune systems. Gu et al. [3] proposed an architecture of combining multiagent systems (MAS) and dendritic cells. Bauer et al. proposed an agent-based model for immune systems [4]. Hamer et al. utilized concepts from artificial immune systems to computer security [5]

Most AIS research are focused on the development of specialized AIS algorithms inspired by theories such as the negative selection theory [6] or the danger theory [7]. Applying AIS algorithm to IDS can be traced back to [9] [10]. For reviews of related works, see [11]. Liu et al. [12] proposed an active defense model for IDS based on immune multiagents (IMA). Dasgupta [13] proposed an immunity-based IDS framework which applied the multi-agent architecture. These schemes are based on negative selection theory.

However, Burgess [14] suggests that negative selection algorithm cannot meet the fast evolution of newly generated computer virus. On the other hand, he put the emphasis of AIS on an autonomous, distributed feedback and healing mechanism. Aickelin et al. [9][15] present the first in-depth discussion on the application of danger theory to intrusion detection. Greensmith et al. [16] employed dendritic cells (DCs) within AIS which coordinated T-cell immune responses. Kim et al. [17] proposed "CARDINAL" which embedded T-cell process within the danger-theory-based AIS. Sarafijanović and Boudec [18] proposed the concept of virtual thymus (VT).

1.2 The Motivation of Emerging AIS and MAS to IDS

Intrusion detection systems, whether utilizing statistical analysis, feature analysis or data mining, have limitations such as self-adaptation, robustness and effective communications. IDSs cannot respond to unknown attacks as they lack self-adaptations. Without robustness, components of IDSs will be isolated each other [12]. On the other hand, if there is no effective communications among components of IDSs, the early warning and response mechanism cannot be established.

AIS can be contributed to the improvements of self-adaptation, moreover, the diversity and memory mechanism of the IDSs. On the other hand, robustness and communications can be improved by mechanisms of multiagent systems. The logical structure of agents with internal reasoning mechanism can effectively solve problems arisen from complicated environment such as complex networks.

The motivation of this research is the following. Can agent-based information security systems learn themselves to effectively determine whether the abnormality is "actually" incurred by some malicious attacks. This paradigm is very similar to the danger theory proposed in the immunology [7]. It leads to the immunity-based multiagent system (IBMAS), which is a promising research topics these days [19].

1.3   Contributions of This Research

This paper combines the advantages of both AIS and MAS to design a better IDS. Several researches related to immunity-based IDS can be further improved by introducing multiagent architectures and utilizing agents' cooperations and communications [20]. This research analyzes some issues of adopting immunity-based IDSs without clarifications of roles of involved systems components. On the other hand, the improvement can be reached by defining role of each agent and the coordinations of these roles.

However, for network infrastructure, there is no entities "on the fly", which means there is no definitive concept for computer hosts "between" two adjacent nodes. This paradigm is quite different from that of human immune system, unless we consider the mobile agents which transit between computer hosts. It is a research area how to identify malicious codes and mobile agents. To be more concentrate is this research, we propose a solution which combines MAS and AIS.

1.4   Approach

The major goal of this paper is to facilitate intelligent agent mechanisms with AIS based on danger theory to improve intrusion detection systems. Such improvements are proposed according to previous researches and architectures such Fu et al. [21] and Kim et al. [22]. Intelligent agents are also required while IDS is considered deploying in the cloud computing environment. For agent-based AIS (ABAIS), these agents embodied the dentritic cell functionality (namely, DC agents) can "detect" danger signal issued by computer hosts being attacked or suspiciously being attacks. While other agents, such as T-cell agent (TC agent), antigen agent (Ag agent) and responding agents (RP agent), communicate one another to improve the efficiency of IDSs. Computer threats generally come from the Internet, which are very similar to those of pathogens to our bodies. The central challenge with computer security is how to discern malicious activities from benign ones. Our intelligent agent-based model majorally consists of the cooperation of DC agents in the innate immune system and TC agents in the adaptive immune system. This dual detective mechanism, where DC agent detects the behavioral information (i.e. signal) caused by an antigen and TC agent detects system call (i.e. antigen), can decrease false positive rate. For the learning process of intelligent system, self-organizing feature map (SOM) network effectively clusters the input (normal) network vectors, while maintaining the topological structure of the input space. ABIDS enables tunable and adaptable threshold values to determine danger signals.

1.5   Structure of This Chapter

The arrangement of this chapter is as follows. In section 2, preliminary knowledge such as AIS, danger theory, intelligent agents system and intrusion detections are introduced. In section 3, ABIDS model inspired by the danger theory is discussed.

Simulations of algorithms related to ABIDS based on different scenarios will be given in section 4. Further analysis of ABIDS will be discussed in section 5.

2   Preliminary Knowledge

2.1   Intrusion Detection Systems (IDS)

Intrusion detection systems (IDS) focus on exploiting attacks, or attempted attacks, on networks and systems in order to take effective measures based on the system security policies, if abnormal patterns or unauthorized access is being suspected. However, there are two potential mistakes by IDS, namely, false positive error (FPE) and false negative error (FNR). For FPE (FNE), a pattern is mistakenly determined as abnormal (normal).

IDSs are used to help protect computer systems. The main goal of IDSs is to detect unauthorized use, misuse and abuse of computer systems by both systems insiders and external intruders [11]. An IDS can be a device or software application that monitors network and/or system activities for malicious activities or policy violations and produces reports to a Security Operating Center (SOC). Kim and Bentley suggested IDSs should satisfy the following seven requirements: robustness, configurability, extendibility, scalability, adaptability, global analysis, and efficiency, for more details see [22][24].

There are two types of IDSs, host-based and network-based intrusion detection system, respectively [23].

Network intrusion detection system (NIDS). It is a platform that identifies intrusions by examining network packets and monitors multiple hosts. Network intrusion detection systems gain access to network packets by connecting to a network hub. Sensors are located at choke points in the network to be monitored; they capture all network packets and analyze the content of individual packets for malicious traffic.

Host-based intrusion detection system (HIDS). It consists of an agent on a host that identifies intrusions by analyzing system calls, application logs, file-system modifications and other host activities and state. Sensors usually consist of a software agent.

2.2   Human Immune System (HIS) and Artificial Immune Systems

Human immune system (HIS) consists of antibodies and lymphocytes, which include varied T-cells and B-cells. HIS uses a large number of highly specific B- and T-cells to recognize antigens. Only B-cells secrete antibodies. Clonal selection theory explains the details of antibody secretion specific to an antigen where T-cells help regulating. The binding between antigen and specific lymphocytes trigger the proliferation from immature lymphocytes to mature one, and the secretion of antibodies. HIS must interact not only with the nonself from the outer world, but also the self from the internal world.

HIS can be categorized as innate and adaptive immune system. The innate immune system is characterized by three roles, namely, host defense in the early stages of infection, induction of the adaptive immune response and determination of the type of adaptive response through antigenic presenting cells (APCs) [25]. On the other hand, the main characteristics of the adaptive immune system are recognitions of pathogens.

Dendritic Cells. Dendritic cell (DC) is a vital link between the innate and adaptive immune system which provides the initial detections of pathogenic invaders. DC is also an APC which captures antigen proteins from the surrounding area and processes it by ingesting and digesting the antigen. DCs are also a part of innate immune system; once activated, they migrate to the lymphoid tissues where they interact with T- and B-cells to initiate the adaptive immune response. Moreover, adaptive immune response is "orchestrated" by DCs.

DCs are the first defense line for HIS which will arrive at the locations where antigens intrude and then swallow the latter to the pieces. These pieces will be attached to APCs and presented to the T-cells. DCs can also be regarded as the commanders for HIS; they can combine the danger and safe signal information to decide if the tissue environment is in distress or is functioning normally. DCs will influence the differentiation of T-cells by releasing particular cytokines. In other words, DCs drive the T-cell to react to the antigen in an appropriate manner.

T-Cell. T-cells coordinate hosts not only by way of Lymph nodes, but also by periphery and Lymph nodes within each host. DCs will influence the differentiation of T-cells by releasing particular cytokines. In other words, DCs drive the T-cell to react to the antigen in an appropriate manner.

Naive T-cells are those have survived the negative and positive selection processes within the thymus, and have migrated to circulation system between the blood and lymphoid organs where they wait antigen presentation by DCs. Naive T-cells reach an activated state when the T-cell receptor (TCR) on their surfaces binds to specific molecules, namely, antigen-peptide-Major Histocompatibility Complex (MHC), on the surfaces of DCs', and costimulatory molecules are sufficiently upregulated on the surface of the DCs to show the degree of danger signals. Activated T-cells proliferate and their clones will differentiate into other cells such as helper T-cells and cytotoxic T-cells (CTL). Differentiation statuses of T-cells play several roles in immunity mechanisms and tolerance in the HIS.

Artificial Immune Systems (AIS), based on HIS, have been applied to anomaly detections [3][12][21][26][27][28]. AISs have been developed according to negative selection algorithm and clonal selection algorithm which are based on the classical self-nonself theory; nonselfs are entities which are not part of human organisms [29]. This so-called self-nonself classification theory had been challenged while failing to explain several immunological phenomena. Some alternative theories have been proposed, for example, the danger theory (DT). DT postulates that the human immune systems respond to the presence of molecules known as danger signals, which are released as results of unnatural cell deaths.

2.3 Computer Security and Immune System

Computer security is composed of processes to prevent malicious programs such as computer virus, internet worms, etc. to invade computer hosts. An important aspect is the following: a computer security system should protect a host or network of hosts from unauthorized intruders, which is analogous in functionality to the immune system protecting the body from invasions by foreign pathogens. Therefore, it is a straightforward consideration to adopt human immune mechanism to computer security.

For computer security, one important aspect learned from immunology is the following: a computer security system should protect a host or network of hosts from unauthorized intruders, which is analogous in functionality to the immune system protecting the body from invasions by foreign pathogens. For example, anti-virus software has recently adopted some features analogous to the innate immune system, which can detect malicious patterns. However, most commercial products do not yet have the adaptive immune system's ability to address novel threats. According to [30], the central challenge with computer security is to determine the difference between normal and potential harmful activities. IT systems are getting larger and more complex, which invokes the needs of developing automated and adaptive defense systems. One promising solution is to acquire AIS.

In general, HIS maintains both low false positive and false negative errors. The former guarantees our immune system can correctly "recognize" harmful pathogens, while the latter for harmless pathogens. This leads to the motivation of AIS-based intrusion detection systems. For example, MHC stimulates antigen presenting cell (APC) to activate, which helps lymphocyte cells identify antigens. Boukerche et al. [31] proposed a mapping between computer security and HIS.

Fig. 1 illustrates a basic architecture of secure computer network. some SOC issues threat profile for each host computer to determine whether a network packet is malicious. It also collects data from each computer host and monitors network behaviors within its security domain. On the other hand, proxy server is in charge of controlling inward and outward network traffic of each computer host. It can be a firewall, an application gateway, or any security gateway. Fig. 1 also suggests an architecture of distributed IDS, which will be discussed later.

For efficient evaluations of IDS, Kim et al. [28] proposed three conditions for an "intelligent" IDS.

1. Optimize the number of peer hosts polled.
2. Types of system response should be determined by attack severity and certainty
3. For performing adequate magnitudes of responses, both local and peer information needs to be taken into account.

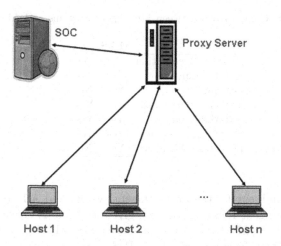

Fig. 1. Architecture of Secure Information System

2.4 Danger Theory

Matzinger [7] proposed the Danger Theory (DT), which has become more popular among immunologists in recent years for the development of peripheral tolerance (tolerance to agents outside of the host). DT states that the immune system will only respond when damage is indicated and is actively suppressed otherwise. It proposes that APCs, (in particular, DCs), have danger signal receptors (DSR) which recognize signals sent out by distressed or damaged cells. These signals inform the immune systems to initiate immune responses. APCs are activated via the danger signals. These activated APCs will be able to provide the necessary signals to the T-cells (more precisely, T-helper cells) which control the adaptive immune response.

Danger signals are generated by ordinary cells of the body that have been injured due to attacks by pathogens. These signals are detected by DCs, which have three modes of operations: immature, semi-mature and mature. In the DC's immature state, it collects antigens along with safe and danger signals from its local environment. DC is able to integrate these signals to decide whether the environment is safe or dangerous. If it is safe, DC becomes semi-matured. Upon presenting antigens to T-cells, DC will cause T-cells to "tolerate". If it is dangerous, DC becomes matured and causes T-cells to become reactive on antigen-presentations.

2.5 Relationship between Multiagent System and AIS

Agent is an entity that has the ability of consciousness, solving problem, self-learning and adapting to the environment, which is very similar to immune systems in functionalities. King et al. [8] proposed an architecture intelligent agents based on AIS. Table 1 summaries such similarities between AIS and MAS.

Table 1. Comparisons between Multiagent Systems and AIS

System	MAS	AIS
Diversity Generation	Yes	Yes
Self Tolerance	Yes	No
Learning	Yes	No
SNS Act. Threshold	Yes	Yes
Self Maintenance	Yes	Yes
Short-term Memory	Yes	Yes
Long-term Memory	Yes	No

On the other hand, similarities of multiagent system and AIS, based on perspectives from intelligent IDSs [22][24], are given as Table 2. According to this table, AIS-based MAS (ABMAS) and Agent-based AIS (ABAIS) are two (loosely coupled) categories for designing intelligent IDS.

Table 2. Comparisons of AIS and Generic MAS

Property	AIS	MAS
Robustness	Yes	No
Configurability	Yes	Yes
Extendibility	No	Yes
Scalability	No	Yes
Adaptability	Yes	No
Global Analysis	Yes	No
Efficiency	Yes	No

Agent-Based Models (ABM). ABM has inspired significant interest as agent-based language is very similar to that of nature. According to [32], ABM is an appropriate method for studying immunology. As computers became more powerful and less expensive, the ABM became a practical method for studying complex systems such as the immune system. The interaction between various types of agents is a criterion to evaluate a multiagent system. For example, "Autonomous Agents for Intrusion Detection" (AAFID) is the first agent-based IDS proposed by Purdue University [33].

To have the agents learn, we may utilize methodologies derived from AIS such as (immune) response attributes of specificity, diversity, memory and self/non-self recognition. Functionalities of the biological immune system such as content addressable memory and adaptation, are identified for use in intelligent agents.

AIS-Based Multiagent Systems (ABMAS). MAS is suitable for task allocation in heterogeneous computing environment, which become a major characteristics for Internet nowadays, in particular for cloud computing. Adaptiveness is a challenge and also an important feature for multiagent system to interact with the environment. Three major stages for ABMAS inspired by the clonal selection theory are diversity generation, self-maintenance and memory of nonself. The last two properties define the adaptiveness of the ABMAS. These steps are carried out by agents distributed over the MAS.

Diversity Generation. (Continuous) diversity generation leads to the "adaptation" of ABMAS. Diverse agents with distinct specificity of the receptor and the effector are generated by way of mutations.

Self-Maintenance. Agents are adjusted to be insensitive to known patterns (self) during the developmental phase. Negative selection theory is a central of this phase.

Memory of Nonself. Agents are adjusted to be more sensitive to unknown patterns (nonself) during the working phase.

CARDINAL. CARDINAL (Cooperative Automated Worm Response and Detection Immune Algorithm) is an immunity-based worm detection inspired by T-cell immunity and tolerance [17]. Although it is not an agent-based architecture, CARDINAL proposed an seamless integration between artificial periphery and lymph nodes which emulate the functionalities of human immune systems, in particular, those of the innate and adaptive immune systems. Mechanisms of T-cell immunity and tolerance provide intelligence of performing criterion of intelligent IDS, see Fig. 2.

CARDINAL also adopts concepts from danger theory. Accordingly, DCs present results of danger signal assessments in three different forms to naive T-cells. The costimulatory signal is increased if a DC detects a severe attack; a strong response to this attack is needed. From HIS, the amount of cytokine IL-12 is increased when a DC detects a severe attack requiring a strong response but

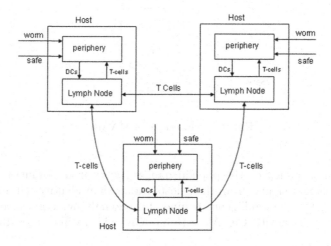

Fig. 2. Architecture of CARDINAL

with a relatively lower certainty. On the other hand, the amount of the cytokine IL-4 is increased when a DC detects a less severe attack which only needs a weak response. CARDINAL does not consider the certainty of attack as a negative effect of a response triggered by a false positive error would be minor. However, it is still an issue worth being studied. One disadvantage of CARDINAL is the assumption that it will target the internet worms with consistent attack signatures. We still think it is a strong hypothesis which should be loosen.

MAAIS. MAAIS (Multiagent-based AIS) was proposed by Fu et al. [21]. The architecture is based on CARDINAL with "concepts" of multiagents. This MAAIS, which consists of two components, namely agents and server processes, provides agent-based anomaly detection functionality. Agents monitor their corresponding hosts; while servers evaluate to select suitable strategies ro respond according to immune response mechanism of immune responses, see Fig. 3. However, [21] did not designate roles of agents specifically. For further improvements, agents may be composed of diverse APCs and various types of artificial T-cells.

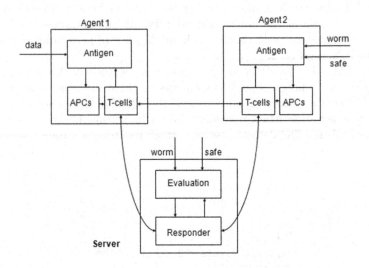

Fig. 3. Architecture of MAAIS

MAAIS is also adopted the popular danger theory from immunology. APCs are sensing danger signals from their hosts rather than identify self and nonself network packets. The intelligence behind this anomaly detection mechanism is abstracted from Dendritic Cell Algorithm (DCA) which will be introduced later.

3   Intrusion Detection Mechanism Based on ABAIS

According to [21][24][31], we propose improved IDS based on MAAIS. Different from traditional self-nonself paradigm for immune systems, our IDS will first detect danger signals emitted by computer hosts. These danger signals are based on some security threat profile, which defines by system calls generated by running processes. According to [27][31], threat profile may be composed of excessive CPU, memory load at the host, bandwidth saturation and high connection number of the host, etc. This threat profile defines what the "event" is. For example, the unknown antigen invoking malicious behaviors such as file deletion, information leakage, etc. In this IDS, several agents are generated which can communicate each other to emulate functionalities extracted from DT-inspired AIS.

3.1   Antigen

For most network detecting systems (for example, those adopting self-nonself paradigm), an antigen is defined as an information vector extracted from network packet. The extracting rule can be very different for each network detecting systems. Moreover, antigens are binary strings extracted from the IP packets, which include IP address, port number, protocol types, etc. However, the antigen defined at our IDS is related to system calls rather than network packets.

A system call is the way how a program requests a service from an operating system's kernel. System calls provide the interface between a process and the operating system. Each process invokes some system calls. The more active the process, the more system calls it makes. Each system call is captured and converted into an antigen attribute [34]. For example, the latter can be represented by CPU usage, memory load at the host, bandwidth saturation and connection number of the host

3.2   Intelligent Agents in AIS

Our ABIDS, which is different from, for example, that of [25], is based on the danger signal rather than self-nonself paradigm. Therefore, we design a MAS with antigen agents, DC agents, T-Cell agents and Responding agents to perform functionalities of IDSs'.

Antigen Agent (Ag Agent). Antigens are profiles of input data such as IP packet, which includes IP address, port number, protocol type, and network connection, etc. Ag agent simply parses input data into format of antigens then sends them to the DC agents.

Antigen agents, which are installed at computer hosts, represent data item from the nonself dataset. They extract and record selected attributes from these data items. As one Ag agent samples multiple times, each antigen agent randomly selects certain amount of DC agents and sends those DC agents a picked message when a nonself antigen appears.

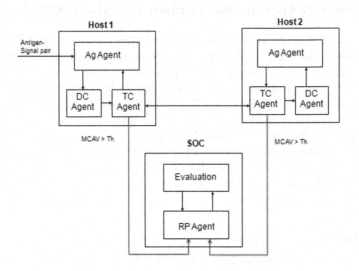

Fig. 4. Architecture of ABIDS

Dendritic Cell Agent (DC Agent). In order to determine whether a malicious behavior is taking place, IDS needs to analyze the input data, i.e., antigens. DC agents are the kernel of the ABAIS which are complex compared to other agents; they are also installed and distributed at each computer hosts. When an Ag agent issues a picked signal, DC agent will evaluate the risk state facing by the host by calculating the danger value (DV) and analyze the signal corresponding to this antigen. Once the DV exceeds some threshold, it will inform the responding installed at the SOC.

Similar to nature DC, each DC agent has three stages, namely immature, semimature and mature. DC agents are started from the immature stages. When a picked signal issued from the Ag agent, DC agent executes data processing function such as the DCA. When a DC agent is at either semimature state or nature state, it returns the mature context to the Ag agent.

T-Cell Agent (TC Agent). TC agents are also installed at each computer hosts. They are activated by the signals from DC agents when the DVs exceed thresholds. Each TC agent has three numerical values associated with it; these represent the accumulated certainties and severities of attack: *T-cell activation threshold, Th1 activation threshold, Th2 activation threshold.* These TC agents will communicate with each other to update these numerical values. There are two perspectives of this agent communications.

- For TC agent issuing warning signal of malicious act to the corresponding antigen, it also informs TC agents "nearby" by exchanging its (three) numerical values.

– For TC agent not issuing warning signal of malicious act for this antigen, it simply updates its numerical values according "nearby" TC agents which have issued warning signals.

This paradigm of TC agents is very similar to that of ensemble neural network (NN), where training NN can be more precise by way of combination of a number of individual networks trained on the same task.

Responding Agent (RP Agent). Ag agents, DC agents and TC agents are co-ordinating one another to perform immune responses. After DV exceeds threshold value, TC agents will inform RP agent, which is installed at some SOC. Therefore it has all these resources to compare the output category with the original category of each antigen, to calculate the overall true positive or accuracy of the virus detections. It represents that an infected host is detected; RP agents will activate some control measure to such malicious action. Two measures are considered [27]:

1. Reporting to the SOC or security manager, for example patches downloaded, activate relevant anti-virus software on this infected host and removes virus.
2. Disruption of intrusion, discards a suspicious packets, kill the related process, cut-off infected sub-network. These can prevent large-scale spreading of computer viruses, in particular internet worms, which have high spreading rate by their natures.

Table 3 is the comparisons for goals and services fulfilled by these agents.

Table 3. Agents, Goals and Services

Agent	Goals	Services
Ag Agent	Data Parsing	antigen label and signal association
DC Agent	Detect malicious codes or network attacks at the input sources	MCAV estimation, update activation threshold
TC Agent	Categorize nonself antigen	Identify the malicious antigen
RP Agent	Reduce the malicious antigen	Implement the system response

3.3 Threshold Values of Intrusion Behavior

Functionalities of the IDS based on ABAIS are heavily depended on determining whether signals issued by computer hosts are dangerous; that is, IDS should have high positive rate of detecting "real" intrusion behaviors. In details, there is an effective threshold values for signals to be most likely intrusion signs. We adopt the Self-Organizing Feature Map (SOM) network proposed as a basis of threshold value generation.

Let $\{X_i\}_{i=1}^{N}$ be a collection of N testing network vectors consisting of network information such as (attack) starting time, (attack) time duration, protocol ID, source port, destination port, source address, destination address and attack profile, etc. Let F be a fixed SOM with input vectors, whose competitive layer is a matrix. In particular, this matrix is tunable by the SOC if necessary. Then the SOM network is trained by these input vectors X_i; the latter can be classified into several clusters afterwards. These clusters can help establish the baseline of normal network behaviors, namely, the threshold values of intrusion behaviors. When a network vector falls into some deviated classes, intrusion signal is issued by some intelligent agent such as DC agents. The following figure is a graphic representation of threshold values generation.

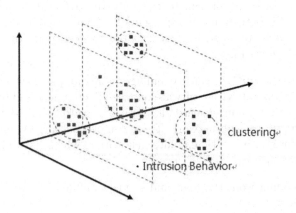

Fig. 5. Intrusion behavior defined by SOM Network

3.4   Agent-Based Dendritic Cell Algorithm (ABDCA)

DCA is an AIS algorithm which is particularly developed for anomaly detection according to the above DCs' functionalities and characteristics [34][35]. It provides information how anomalous a group of antigens is. This is achieved through a generation of an anomaly coefficients, namely, *mature context antigenic value (MCAV)*. It is believed that a DC is better performed by agent technology while considering its adoption to network environment. This antigen-plus-context information is passed on to a class of responder cells, termed T-cells.

DC agents in our ABIDS will evaluate antigens and corresponding signals according to the DCA to determine whether antigens are malicious nor not. The "context" means the classification identifier for an antigen. If an antigen is collected in an environment of danger, the context of this antigen is marked as anomalous and such antigen collected by the DC agent is potentially an intruder. While in the immature state (namely, initial phase), DC agent performs the following three functions:

1. Antigen Sampling: DC agents collect antigens from some external sources (in this case, from computer hosts) and places these antigens in its own antigen storage.
2. Update input signals: DC agent collect values of all input signals present in the signal storage area.
3. Calculate interim output signals: each DC agent calculates its three temporary output signal values from the received input signals, then derives its final output signal.

We observe that multiple DCs present multiple copies of the same antigen type for invoking an immune response; it leads to an error-tolerance immune system as a single DC is far from stimulating a false positive error. The ABDCA is listed as follows.

Algorithm 1. Agent-based DCA (ABDCA)

---

input : Antigens and signals
output: Antigen context (0 for safe/1 for danger)
Initialize DC agents (Immature state);
for each DC_agent_i **do**
   get antigen (Ag);
   store antigen;
   get input_signal;
   calculate output_signal_i;
   if output_signal_i > Activation_Threshold_i **then**
     Ag_context is assigned as 0;
     State of DC_agent_i="semi-mature";
   else
     Ag_context is assigned as 1;
     state of DC_agent_i="mature";
   end if
end for
update cumulative output_signals;

Three temporary output signals are PAMP signal, danger signal and safe signal. According to Greensmith et al. [34], the system should respond with a very high rate of false positives by switching the PAMP and safe signal. The definitions of PAMP, danger and safe signals, by way of parameters of computer hosts such as CPU usage, memory load, network connection (number) and bandwidth saturation, are as follows.

1. PAMP Signal: Network Connection > th_netconnection AND bandwidth Saturation > th_bandsaturation.
2. Danger Signal: CPU Usage > th_cpuload OR Memory Load > th_memoryload.
3. Safe Signal: All parameters < th_parameter.

Where "th_parameter" represents the threshold value of the parameter. Now the computation of output signals by three temporary output values is the following.

$$output\_signal = \frac{W_P \cdot C_P + W_D \cdot C_D + W_S \cdot C_S}{W_P + W_D + W_S} \tag{1}$$

According to empirical experiments [35], the weights for the output signal is $W_P = 2$, $W_D = 1$, $W_S = 1$. C_P, C_D, C_S represent PAMP, danger and safe signal, respectively. The principle for these weights is the following: PAMP signal will decide whether the danger signal is a "really" harmful one; whether output signal is harmful or nor is defined by the Table 4.

Table 4. Definitions of Output Signal

Output Signal	C_P	C_D	C_S
Normality	0	0	1
Harmless Abnormality	0	1	0
Harm Abnormality	1	1	0

Threat Profile. One issue of ABIDS is the "baseline" for determining abnormality of network packets. Such baseline, namely the *threat profile*, provides intelligent determination of attack types of network packets. It can be determined by three factors, namely, attack severity (S), certainty (C) and the length of attack time (T) [21]. There are different aspects of estimating S, C and T. From network detection viewpoints, these factors are functions of Ag agent attribute (CPU_usage, memory_load, bandwidth_saturation, connection_ numbers). S, C, and T are normalized, namely, $S, C, T \in [0,1]$. Threat profile is a vector $\langle W_S, W_C, W_T \rangle$, where W_S, W_C, W_T are weighted factors of S, C, T respectively.

Event-Driven Architecture. If a network packet is not equal to a pattern or signature that the system has already stored, the current pattern-based architecture does not detect it as abnormal. This happens for a virus with mutation capability. In general, IDSs cannot detect such transformations. Such inflexibility leads to the consideration of event-driven architecture for IDSs. One important factor for such improvement is the **certainty**. In order to improve the false negative rate, we have to discard all detected events whose attack certainties are below a threshold value. The following is a definition of an attack.

Definition 1. *An incident for a network domain is called an "attack", if its antigen attribute is greater than its corresponding threshold value.*

Technologically, incident of attack can be determined by some rule-based methodologies such as the attack graph methodology based on system call sequence. In order to reduce both the false positive and false negative rates, the decent ABIDS should cooperate both the danger theory and attack graph. The paradigm of the latter is out of scope of this research.

Algorithm: Agent-Based IDS. Now according to ABDCA, we propose an algorithm describing agent-based IDS. This algorithm is also illustrated as Fig. 6. ABDCA can provide information not only a network packet but also a group of network packets is anomalous or not. This is achieved by the generation of an anomaly coefficient value, namely MCAV.

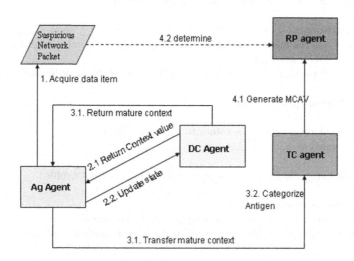

Fig. 6. Diagrammatic Illustration of Agent-based intrusion detection system (index numbers are referred to Algorithm 2

For the step 4 of the Algorithm 2, each nonself antigen gets a binary string of mature contexts from every selected DC agent installed at corresponding computer host. MCAV can be calculated through the number of context "1" divided by the number of all contexts. According to [3], it is similar to a voting system, where the antigen is the candidate and the DC agents are the voters. If the context is "1" ("0"), it means the DC agent determines this antigen is malicious (benign). The MCAV is actually the probability of that this antigen is being malicious.

The reactions of RP agents can be multiple according to the SOC's security policies. For example, passive reaction is initiated by sending some alarm signal such as an e-mail to the administrator and a strategy proactive is defined through mobile agents that implement the characteristics of the reaction. According to [31], for example, an applied reaction is defined for each of the following internet protocols such as DNS, FTP, HTTP, POP3 and SMTP. The issue of RP agents is out of the scope of this research.

3.5   Functionalities of ABIDS

Security Response by SOC. SOC plays the central role of security responding mechanism. Once MCAV exceeds some threshold, RP agent installed in the

---

Algorithm 2. Agent-based IDS (ABIDS)

Initialization: Some network packets are suspicious as malicious.

Input: Antigen-Signal pair

Output: Antigen Type

1. Data Processing:

1.1 Ag agent extracts an antigen from Antigen-Signal pair.

2. Agent response:

2.1 DC agent returns its context value according to ABDCA; also according to the policy of the SOC if necessary.

2.2 DC agent responds to this antigen by updating its state according to ABDCA.

3. Danger Signal processing:

3.1

if this DC agent returns mature values to Ag agent **then**

 Ag agent transfers it to the TC agent.

end if

3.2 TC agent categorizes such antigen according to the threat profile.

4. MCAV generation:

4.1 TC agent generates MCAV, which is sent to the RP agent.

4.2 RP agent determines if the corresponding antigen is malicious or not.

SOC is activated and makes a comprehensive evaluation for the received danger signal. This evaluation is a crucial factor to mitigate the threat of the whole network. If the evaluation is not good enough, the false positive error will produce the damage equal to the one caused by the attack itself. This observation, coincidentally agreed with the paradigm of danger theory, suggests that the comprehensive evaluation should be depending not only on the danger signal, but also the number of DC agents emitting danger signals.

Let CE represent the comprehensive evaluation, AVE is the average value of DV exceeding threshold value T; n is the number of DC agents emitting danger signals, N is the total number of DC agents. The calculations of CE and AVE are as follows; Table 5 is the suggesting weights W_A and W_n.

$$CE = \frac{W_A \cdot AVE + W_n \cdot (n/N)}{W_A + W_n} \tag{2}$$

$$AVE = \frac{1}{n} \sum_{i=1}^{n} MCAV_i \tag{3}$$

$CE, AVE \in [0, 1]$. There will be a critical task for SOC to define its security measures according to CE values. This issue is also out of our scope here.

Agents' Communications and Coordination. One advantage of ABIDS is the communications and coordination between DC and/or TC agents from

Table 5. Weights related to CE

W_A	W_n
2	1
1	1
1	2

corresponding hosts. The basic idea is as follows. TELL and ASK-based communication permit agents share their internal information, enhancing their performances to respond to intruders [21]. On the other hand, useful knowledge obtained by each agent should be stored in a database. These databases could also be shared by each agent to improve their detection efficiencies. Some suitable mechanism of agent communication between agents can contributed to better performance of ABIDS. Table 6 illustrates the agent communications [5].

Table 6. Agents' Communications

Communication	Initiator	Receiver	Description
eRaiseWarning	Ag Agent	(Selected) DC agents	Notify of a possible malicious attack
eContextReturned	DC Agent	TC Agent	sending MACV
eMaliciousOrNot	DC agent	TC agent	Identify the malicious antigen
eEvidenceResponse	TC Agent	RP agent	Implement the system response according to TC agent's information

4   Simulation

4.1   Malicious Behaviors Determined by ABDCA

According to essence of the danger theory, the advantage of ABIDS is the following: it can determine some nonself-antigenic behavior which is at the verge of normality and abnormality. For example, for those network behaviors with short attacking time. In this section, we simulate several critical cases for this ABIDS. The threshold of S, C and T are the following $S_{th} = 0.50$, $C_{th} = 0.50$ and $T_{th} = 0.5$. The number of computer hosts within this network is 2000. Each host has its fixed vector of weights. Simulation principle is to determine those critical behaviors, which include one extremely high factor with other low factors, or some factors are closing to the corresponding threshold values.

Nonself Antigen with High Severity, Low Certainty and Short Attacking Time. While an antigen induces a computer host with behaviors of high severity but low certainty and short attacking time, it is difficult sometimes to determine if this is a malicious attack. In this case, we consider the threat profile TP=[0.9, 0.01, 0.01] for the ABDCA algorithm. We expect the simulation should be "stable" for reasonable many hosts as Fig. 7. MCAV is around 0.03 and 0.05, which shows that this nonself antigen is normal.

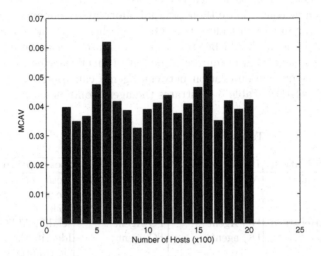

Fig. 7. Average MCAV for an Antigen with High Severity solely

Nonself Antigen with Edge Behavior. While an antigen induces a computer host with edge behavior such that severity, certainty and attacking time are all closing to their threshold values, respectively. It is also difficult to determine if this is a malicious attack or not. In this case, we consider TP=[0.65, 0.65, 0.65]. The simulation also show a stable result (see Fig. 8); average MCAV is equal to 1, which indicates that this nonself antigen is harmfully abnormal.

Nonself Antigen with Short Time Attack. In this simulation, an antigen causes computer hosts with both high severity and certainty, but relative medium attacking time. It is difficult to determine if this is a malicious attack or not. In this case, we consider TP=[0.9, 0.9, 0.4]. The simulation show a stable result (see Fig. 9); average MCAV is around 0.5, which indicates that this nonself antigen can be harmfully or harmful abnormal depending on the SOC's security profile.

Antigen with Long Time Attack. Another interesting network behavior is the following: the connection time is relatively long but with medium severity and certainty. The following simulation is for antigen profile $TP = [0.3, 0.6, 0.8]$.

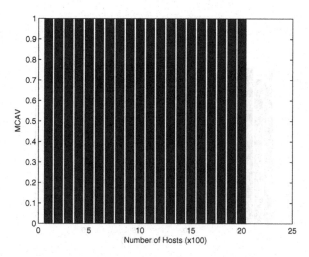

Fig. 8. Average MCAV for an Antigen with Critical Threshold Behavior

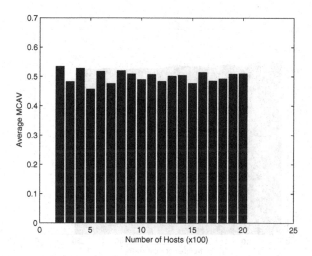

Fig. 9. Average MCAV for an Antigen with High Severity and certainty, but relatively short Attacking Time

According to Fig. 10, the average MCAV is stably around 0.4 and 0.5. This nonself antigen can be harmfully or harmful abnormal depending on the SOC's security profile.

4.2   Simulation with Costimulation Signals

In this subsection, we will concentrate on the actual host behaviors rather than threat profile. These host behaviors (HB) include CPU usage, memory load,

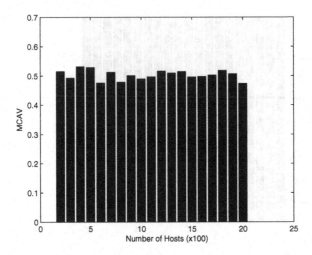

Fig. 10. Average MCAV for an Antigen with Relatively High Attacking Time

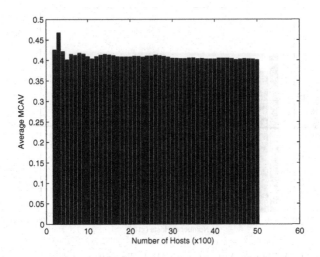

Fig. 11. Average MCAV for an Antigen causing host behavior $HB = [0.95, 0.8, 0.8, 0.3]$

network connection (numbers) and bandwidth saturation. We first consider the following host behavior: $HB = [0.95, 0.8, 0.8, 0.3]$. This is the case where bandwidth is relatively small. According to Fig. 11, the average MCAV is stably around 0.4. On the contrary, we also simulate the similar case only for large bandwidth saturation: $HB = [0.95, 0.8, 0.9, 0.9]$. According to Fig. 12, the average MCAV is stably around 1, which shows this nonself antigen is definitely a malicious one.

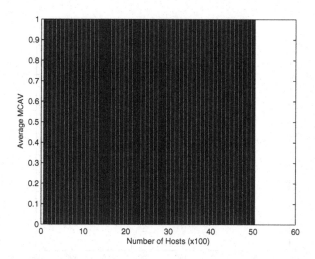

Fig. 12. Average MCAV for an Antigen causing host behavior HB=[0.95, 0.8, 0.9, 0.9]

4.3   Simulation of TC Agents' Coordination

If a TC agent determines its corresponding host is under malicious attack, it will inform other TC agent "nearby". Such communication is direct and does not pass through proxy server or SOC. The TC activation threshold T will be updated according to the following equation.

$$T(t) = T(t-1) + \alpha E(t-1)(E(t-1) - T(t-1)), t = 1, 2, \ldots, t_b \qquad (4)$$

where E is the excitation level of TC, and t_b is the attacking time by this antigen. Fig. 13 is a simulation result where attacks incurs around time 150.

Interaction between Two TC Agents. We consider two computer hosts with TC interactions. TC1 and TC2 are informing each other for confirmed attack from the corresponding antigen. According to Fig. 14, even two TC agents will lead to "more" complex threshold behaviors.

Interaction among Three TC Agents. We consider three computer hosts with TC agents interactions. TC1, TC2 and TC3 are informing each other for confirmed attack from the corresponding antigen. According to Fig. 15, three TC agents will lead to "more" complex threshold behaviors.

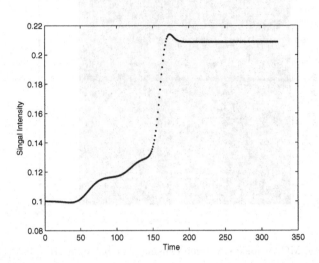

Fig. 13. Activation Threshold of a TC agent within the attacking cycle

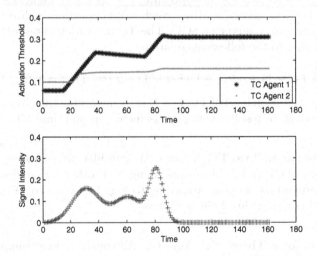

Fig. 14. Activation Thresholds of Two TC agents within the attacking cycle

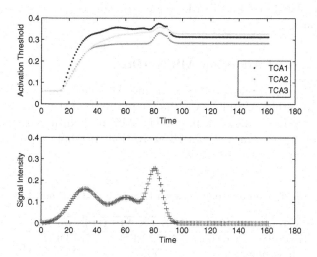

Fig. 15. Activation Thresholds of Three TC agents within the attacking cycle

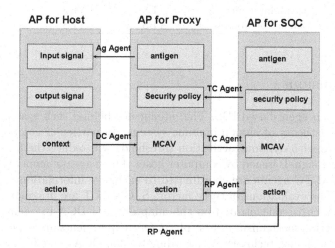

Fig. 16. Agent Interaction for ABIDS

5   Analysis of ABIDS

5.1   Agent Interactions

One advantage of this ABIDS is the feasibility of being adapted into large-scale computer networks such as cloud computing environment, as TC agents will collect information from other computer hosts to update threat profile. Fig. 16 illustrates the agent interactions between SOC, computer hosts and proxy server on their agent platforms. This can be regarded as a security mechanism based on cloud computing.

5.2 Advantages of Adopting AIS to IDS

Table 7 is the comparisons for IDS adopting AIS mechanism such as danger theory. We realize that AIS improves the adaptability of IDS such as self tolerance and learning mechanism. In particular, AIS mechanism can contribute to the updates automation of Self-nonself activation threshold, which is a major concern for current network security.

Table 7. Comparisons between IDS with and without AIS

System	IDS /w AIS	IDS /wo AIS
Diversity Generation	NO NEED	NEEDED
Self Tolerance	YES	NO
Learning Mechanism	GOOD	MEDIUM
Automated updates	**YES**	**NO**
SNS Act.Threshold	**YES**	**NO**
Self Maintenance	YES	YES
Memory of Nonself	YES	YES

6 Conclusions

We propose an agent-based IDS. The intelligence behind such system is based on the danger theory of human immune systems. In particular computations of danger values with dynamic activation thresholds will reduce the false positive rate of danger signals issued by computer hosts. Three agents, namely, Ag agent, DC agent and TC agents are coordinated to exchange information of intrusion detections. The evaluations of three factors S, C, and T are pragmatic issues.

This research is concentrated on the adaptation of DCA to agent-based intrusion detection mechanism in first place. Our experiments are concerned with the feasibility of such technology. According to [36], DCA can effectively reduce both the false positive rate and the false negative rate while applied to detection of port scan-based attacks. Further consideration of importing real network packets for discussing FPR and FNR will be our future goal.

References

1. The Ten Most Critical Web Application Security Vulnerabilities, 2007 update, 2002-2007 OWASP Foundation
2. Forrest, S., Hofmeyr, S.A., Somayaji, A., Longstaff, T.A.: A sense of self for Unix processes. In: Proceedinges of the 1996 IEEE Symposium on Research in Security and Privacy, pp. 120–128. IEEE Computer Society Press (1996)

3. Gu, F., Aickelin, U., Greensmith, J.: An Agent-based classification model. In: 9th European Agent Systems Summer School (EASSS 2007), Durham, UK (2007), http://arxiv.org/ftp/arxiv/papers/0910/0910.2874.pdf
4. Bauer, A., Beauchemin, C., Perelson, A.: Agent-based modeling of host-pathogen systems: the successes and challenges. Information Sciences 179, 1379–1389 (2009)
5. Harmer, P., Williams, P., Gunsch, G., Lamont, G.: An Artificial Immune System Architecture for Computer Security Applications. IEEE Transactions on Evolutionary Computations 65(3), 252–280 (2002)
6. Burnet, F.: The clonal selection theory of acquired immunity. Cambridge Univerisity Press (1995)
7. Matzinger, P.: Tolarance, Danger and the Extended Family. Annual Review in Immunology 12, 991–1045 (1994)
8. King, R., Russ, S., Lambert, A., Reese, D.: An Artificial Immune System Model for Intelligent Agents. Future Generation Computer Systems 17(4), 335–343 (2001)
9. Aickelin, U., Bentley, P., Cayzer, S., Kim, J., McLeod, J.: Danger Theory: The Link between AIS and IDS? In: Timmis, J., Bentley, P.J., Hart, E. (eds.) ICARIS 2003. LNCS, vol. 2787, pp. 147–155. Springer, Heidelberg (2003)
10. Kephart, J.: A Biologically Inspired Immune System for Computers. In: Artificial Life IV Proceedings of the Fourth International Workshop on the Synthesis and Simulation of Living Systems, pp. 130–139. MIT Press
11. Kim, J., Bentley, P., Aickelin, U., Greensmith, J., Tedesco, G., Twycross, J.: Immune system approaches to intrusion detection: a review. Natural Computing 6(4), 413–466 (2007)
12. Liu, S., Li, T., Wang, D., Zhao, K., Gong, X., Hu, X., Xu, C., Liang, G.: Immune Multi-agent Active Defense Model for Network Intrusion. In: Wang, T.-D., Li, X., Chen, S.-H., Wang, X., Abbass, H.A., Iba, H., Chen, G.-L., Yao, X. (eds.) SEAL 2006. LNCS, vol. 4247, pp. 104–111. Springer, Heidelberg (2006)
13. Dasgupta, D.: Immunity-based intrusion detection systems: a general framework. In: Proceeding of the 22nd National Information Systems Security Conference, NISSC (October 1999)
14. Burgess, M.: Computer Immunology. In: Proceedings of the Twelfth Systems Administration Conference (LISA 1998), Boston, Mass, December 6-11 (1998)
15. Aickelin, U., Cayzer, S.: The danger theory and its application to ais. In: Proceedings of the First International Conference on Artificial Immune System (ICARIS 2002), pp. 141–148 (2002)
16. Greensmith, J., Aickelin, J., Cayzer, S.: Detecting danger, The Dendritic Cell Algorithm. Robust Intelligent Systems 12, 89–112 (2008)
17. Kim, J., Wilson, W.O., Aickelin, U., McLeod, J.: Cooperative Automated Worm Response and Detection ImmuNe ALgorithm(CARDINAL) Inspired by T-Cell Immunity and Tolerance. In: Jacob, C., Pilat, M.L., Bentley, P.J., Timmis, J.I. (eds.) ICARIS 2005. LNCS, vol. 3627, pp. 168–181. Springer, Heidelberg (2005)
18. Sarafijanović, S., Boudec, J.Y.: An Artificial Immune SYstem for Misbehavior Detection in Mobile Ad-Hoc Networks with Virtual Thymus, Clustering, Danger Signal and Memory Detectors. International Journal of Unconventional Computing 1, 221–254 (2005)
19. Ishida, Y.: Immunity-based Systems. A Design Perspective. Springer (2004)
20. Weiss, G.: Multiagent Systems, A Modern Approach to Distributed Artificial Intelligence. MIT Press (1999)
21. Fu, H., Yuan, X., Wang, N.: Multi-agents artificial immune system (MAAIS) inspired by danger theory for anomaly detection. In: International Conference on Computational Intelligence and Security Workshops, pp. 570–573 (2007)

22. Kim, J., Bentley, P.: The Artificial Immune Model for network intrusion detection. In: Proceeding of European Congress on Intelligent Techniques and Soft Computing (EUFIT 1999), Aachen, Germany (September 1999)
23. Scarfone, K., Mell, P.: Guide to Intrusion Detection and Prevention Systems (IDPS). Computer Security Resource Center (National Institute of Standards and Technology) (800-94) (February 2007)
24. Kim, J.: Integrating artificial immune algorithms for intrusion detection, Ph.D thesis, University College London (2002)
25. Yeom, K.-W., Park, J.-H.: An artificial immune system model for multi agents based resource discovery in distributed environments. In: Proceedings of the First International Conference on Innovative Computing, Information and Control, pp. 234–239 (2006)
26. Aickelin, U., Bentley, P., Cayzer, S., Kim, J., McLeod, J.: Danger Theory: The Link between AIS and IDS? In: Timmis, J., Bentley, P.J., Hart, E. (eds.) ICARIS 2003. LNCS, vol. 2787, pp. 147–155. Springer, Heidelberg (2003)
27. Zhang, J., Liang, Y.: Integrating Innate and Adaptive Immunity for Worm Detection. In: Second International Workshop on Knowledge Discovery and Data Mining (WKDD 2009), pp. 693–696 (2009)
28. Kim, J., Bentley, P., Aickelin, U., Greensmith, J., Tedesco, G., Twycross, J.: Immune system approaches to intrusion detection-A review. Nature Computing 6, 413–466 (2007)
29. Hofmeyr, S., Forrest, S.: Immunity by Design. In: Proc. of the Genetic and Evolutionary Computation Conference (GECCO), pp. 1289–1296 (1999)
30. Yeom, K.-W.: Immune-inspired Algorithm for Anomaly Detection. In: Nedjah, N., Abraham, A., de Macedo Mourelle, L. (eds.) Computational Intelligence in Information Assurance and Security. SCI, vol. 57, pp. 129–154. Springer, Heidelberg (2007)
31. Boukerche, A., Machado, R., Juca, K., Sobral, J., Motarem, M.: An Agent based and Biological Inspired Real-time Intrusion Detection and Security Model for Computer Network Operations. Computer Communications 20, 2649–2660 (2007)
32. Forrest, S., Beauchemin, C.: Computer immunology. Immunological Reviews 216(1), 176–197 (2007)
33. Spafford, E., Zamboni, D.: Intrusion detection using autonomous agents. Computer Networks 34(4), 547–570 (2000)
34. Greensmith, J., Aickelin, U., Tedesco, G.: Information fusion for anomaly detection with the Dendritic Cell Algorithm. Information Fusion 11, 21–34 (2010)
35. Greensmith, J., Feyereisl, J., Aickelin, U.: The DCA: SOMe Comparison. Evolutionary Intelligence 1(2), 85–112 (2008)
36. Greensmith, J., Aickelin, U., Cayzer, S.: Introducing Dendritic Cells as a Novel Immune-Inspired Algorithm for Anomaly Detection. In: Jacob, C., Pilat, M.L., Bentley, P.J., Timmis, J.I. (eds.) ICARIS 2005. LNCS, vol. 3627, pp. 153–167. Springer, Heidelberg (2005)

Security of E-Commerce Software Systems

Esmiralda Moradian

Department of Software and Computer Systems,
KTH Royal Institute of Technology,
Forum 120, 164 40 Kista, Sweden
moradian@kth.se

Abstract. Cybercrime is costly both for businesses and consumers. Criminals can have different purposes, such as financial winnings, defacement and disruption, which not only cause financial loss but also damage organization's reputation and image. To prevent a number of cybercrimes and simple mistakes, such as not insuring that all traffic into and out of a network pass through firewall, security of e-commerce systems should be considered from the very beginning, i.e. early stage of the e-commerce software development. This is due to software vulnerabilities are a huge security problem. Therefore, to enhance security of e-commerce software, we propose the use of multi-agent system. The research in this paper is focused mainly on the design of agents that provide support to engineers during development process. Moreover, the multi-agent system, presented in this research, supports implementation of patterns and extraction of security information, and provides traceability of security requirements in the engineering process.

Keywords: E-commerce, software system, security, multi-agent system, decision support.

1   Introduction

Software systems do not exist in a vacuum and must, during their lifetime, interact with the users, other systems (both software and hardware), and the environment. Regardless, a huge number of software systems that are used in e-commerce are developed without security consideration. E-commerce software systems can induce security weaknesses and defects, if security is not considered during the software development process.

Ignoring security, commonly, result in vulnerabilities that can be exploited by criminal and malicious users. Malicious users, which often belong to criminal organizations', exercise attacks that target specific selected software of which most provide access to sensitive information. Weaknesses in software may be exploited to gain access to and control of the system, steal sensitive information via the system, and use the system against the owners and users [14]. It is possible due to vulnerabilities that software possesses. Hence, it is essential to reduce vulnerabilities, which are usually the result of defective requirements and design specifications, but also implementation and testing. Software bugs are unknowingly or consciously injected into software

A. Håkansson & R. Hartung (Eds.): *Agent & Multi-Agent Syst. in Distrib. Syst.*, SCI 462 pp. 95–103.
DOI: 10.1007/978-3-642-35208-9_5   © Springer-Verlag Berlin Heidelberg 2013

by developers [7]. Security issues in software systems, nowadays, have been given attention due to the severe impacts software systems have on the human society, including e-commerce, governmental, military, financial, health, telecom, and transport sectors. Since security is not considered during development process, a large number of these systems, unfortunately, contain vulnerabilities, and are not resistant to attacks.

Nowadays, agent technologies are used in information and communication systems in order to provide management, search, monitoring, etc. Multi-agent systems have, also, been used for monitoring and logging purposes as well as for network security. To consider security during development and to be able to build secure software systems, engineers need support provided not only by security specialists but also by automated tools. However, to our best knowledge, such support does not exist. In this paper, we propose multi-agent system that can support engineers during the development process. Presence of semi-automated multi-agent system is necessary due to several reasons, such as: monitor decisions and activities; search for security measures and mechanisms; perform checks; and provide advices and feedbacks.

In this research, we focused mainly on the design of agents that provide support to engineers during development process. Moreover, the multi-agent system provides monitoring of the software development process, traceability of security requirements, and gives security advices in a form of checklists. Further, the proposed multi-agent system supports implementation of ontology-based patterns by extracting security information.

2   Related Work

Jennings et al. [5] in the paper entitled "Autonomous Agents for Business process management" suggest using intelligent agents to manage business processes. The authors describe in their paper the motivation, conceptualization, design and implementation of an agent-based business process management system. In the proposed system, responsibility for enacting various components of the business process is delegated to a number of autonomous problem-solving agents. To enact their role, these agents typically interact and negotiate with other agents to coordinate their actions and to buy in the services they require.

Fasli [3] discusses the use of intelligent agents in e-commerce and highlights risks that emanate from stealing information.

Mařík and McFarlane [6] discuss the need of agent-based solutions for enterprise system to provide intelligent decision-making, manufacturing control, monitoring, and transportation purposes. The authors state that intelligent agent-based technology has been applied to solve different problems like distributed order pre-processing, production planning processes and financial management and billing [6].

In our research, we propose the use of agents in multi-agent system to support engineers during software development process in order to facilitate development of more secure software for e-commerce use.

3   Essence of Software Security Engineering

Systems are often built, operated, and maintained by different groups or organizations. "Software flaws and defects can cause software to behave incorrectly and unpredictably, even when it is used purely as its designers intended" [4]. Vulnerabilities in software that are introduced by mistake or poor practices are a serious problem today. Since software controls the organizations systems, the loss of information, as well as financial loss, depends to a large extent, on how insecure software is [4, 7, 16]. Software development process offers opportunities to insert malicious code and to unintentionally design and build software with exploitable weaknesses.

By using firewalls and/or Intrusion Detection Systems (IDS), e-commerce organizations are often trying to protect themselves from attackers but are unaware that their assets are exposed even through firewalls and IDS's. E-commerce software can be exposed in many ways; therefore, developers of e-commerce systems need to decide how the software should react in different situations in a defined environment. However, it is difficult to perceive progress and traceability in software development [17]. Therefore, security-enhanced processes and practices—and the skilled people to perform them—are required to build software that can be trusted not to increase risk exposure [4].

"Software security has as its primary goals three aspects, the preservation of the confidentiality, integrity, and availability of the information assets and resources that the software creates, stores, processes, or transmits including the executing programs themselves" [2]. Security criteria should be included in each SDLC phase's input and output checkpoints [4].

Unfortunately, security solutions are often isolated from the system functionality, and can be inadequate to the stakeholders' requirements. Hence, multi-agent system that is able to provide support for developers and enhance traceability of security requirements throughout the development lifecycle is highly needed.

Requirements traceability is necessary in order to ensure that requirements are not lost in the design or implementation phases [4]. Pfleeger [15] states that capturing requirements is one of the critical parts of development process, which affects system development during all other phases. We argue that threats, attacks, and vulnerabilities are imperative factors from which security requirements are derived. Therefore, traceability between these factors in relationship with stakeholders' requirements, laws and regulations should result in definition of the security requirements.

We propose the multi-agent system that supports stakeholders, developers and managers during the engineering lifecycle. The purpose of the multi-agent system is to provide monitoring over the development lifecycle, verify and validate security requirements, and provide advices regarding security activities and security controls in a form of checklists.

4   Multi-Agent System

The proposed multi-agent system is a web-based system that can operate both internally, e.g. within the organization's network, and externally, in distributed networks, since data and information can reside in distributed environment.

The environment that agents can work within is cooperative, accessible, episodic, deterministic, dynamic and discrete. In the cooperative environment, communication between the agents takes place. Accessible environment implies that agents have access to the information and knowledge needed to perform a task. The environment is divided into atomic episodes, where each episode has an agent that performs a single task. Deterministic means that next state of the environment is determined by the current state and the action that is being executed by an agent. Dynamic environment refers to the environment that can change. Discrete environment can have a finite number of states; it also can have a discrete set of perceptions and actions. Dynamic environment refers to the environment that can change. [9,10]

The agents are communicative, mobile, cooperative, goal-oriented, autonomous, adaptive, and reactive. The agents are mobile, and can move between different locations over the networks while searching for components and services [11].

The system concerns with how agents cooperate to achieve goals, i.e., requests from users, and what is required of each individual agent in order to accomplish the goals.

Multi-agent system consists of different modules, such as, interface and authentication module, search module, and match and check module. Two types of agents are used: meta-agents and software agents. The agents in multi-agent system are organized in hierarchy, which consists of meta-agents and ground level software agents. Meta-agents operate at macro-level, while software agents operate at micro-level. The meta-agents are autonomous. It means that the meta-agents are able to take decision to satisfy their objectives [19]. Meta-agent makes a choice of best alternative regarding request from the user [8]. Implemented alternative leads to outputs and results that should satisfy the predefined goals [18]. Before the choice can be made, each alternative must be evaluated in terms of the extent to which they satisfy the objectives, i.e. defined goals.

In the proposed system some of the concepts, such as hierarchy, roles, responsibilities, and permissions are adopted from Gaia methodology [1].

Each role, that the agent possesses, is associated with a service (function). Each service has an input, an output, pre-condition, and post-condition [14]. For example, as input 'GetDocument' function can take keywords (for example, asset pattern, threat pattern) and a security level value and compare input tags to document tags. For instance, pre-condition and post-condition for 'GetDocument' function is that the knowledge base must not be empty (knowledge base empty=false) [14].

A role is defined by following attributes: responsibilities, permissions, and activities. A role also has the ability to generate information, to monitor and log events. Responsibilities are divided into two properties, such as satisfaction and security. Satisfaction defines states where an agent fulfills the goal. Satisfaction expression are activities that define an action the agent can perform, for example, SearchAgent = search [14]. The agents work with one task at time. To increase efficiency and shorten search time, software agents, in our work, execute in parallel, i.e., the group of software agents can perform one or more tasks to one or more destinations [9].

Every activity corresponds to a security property. Security property states that the system is monitored and security properties of the multi-agent system, such as confidentiality, integrity, availability, accountability, and non-repudiation, are satisfied.

Agents are assigned permissions due to responsibilities. The principle of least privilege is applied [9, 11], which implies that an agent must be able to access, i.e., read, write, execute, and/or generate only the resources (information resources or knowledge the agent possess) that are necessary for its legitimate purpose, i.e., execution and fulfillment of the assigned task (goal). Communication pathways define communication between agents, which can be unidirectional (a→b) or bidirectional (c ↔ d). The process is initiated by user request.

Each agent has allocated specific roles, which for meta-agents can be any of following: InterfaceAgent, AuthenticatorAgent, Management and Coordination Agent, ControlAgent, and MatchAgent. Ground level software agent is assigned a SearchAgent role. The IntefaceAgent processes requests from developers where a search request can contain terminology, as well as classes and/or properties. The AuthenticatorAgent authenticates the user and checks access rights. The ManagementAgent assigns tasks, coordinates the search, and manages software agents. Software agents can retrieve ontology-based patterns from local knowledge repository, as well as data from databases and repositories, for example, vulnerability databases and control catalogues. The ManagementAgent merges the results from the SoftwareAgents, which means that a meta-agent can compare and map different ontologies [13]. The process is as follows: InterfaceAgent receives a request from the user. The request (message) is validated if all constraints satisfied and passed to the ManagementAgent. ManagementAgent manages and coordinates the activities of SoftwareAgents. ManagementAgent also assigns the task to the Software Agents, to search and retrieve ontology patterns from the repository. The user in MAS is defined as a human agent (HA). Human agent roll can be assigned to stakeholders, managers, and engineers, i.e., security specialists, requirements engineers, architects, designers, programmers, and testers.

To the human agent, Interface agent is acting as an interface. A Role Schema of the InterfaceAgent is presented in Table 1:

Table 1. Role Schema Interface Agent. Source [14]

Role Schema: InterfaceAgent
Description Receives request from the user (HA), passes request to ManagementCoordinationAgent, and returns response.
Function and Activities: 　　　CheckForRequest, AuthenticateUser, CheckAccessRights, PassRequest, 　　　InformUser. Permissions: 　　　　　　　read, execute　　　supplied userInformation　//login information 　　　　　　　generates　　　　　checklist Responsibilities: 　Satisfaction　　　　InterfaceAgent = (IdentifyUser. PassRequest. GenerateCheck- list) 　　　　　　　　　IdentifyUser = (AuthenticateUser. CheckAccessRights) 　　　　　　　　　GenerateChecklist = (ProduceChecklist. InformUser) Security: 　　• UserInformation = bad ⇒ login =nil 　　• LoginAttempt (monitorEvent, logEvent)

In table 1, role, functions and activities of the InterfaceAgent are illustrated. The activities are as follows: AuthenticateUser, CheckAccessRights, PassRequest, and InformUser. The agent possesses following permissions: read and execute the login information, provided by the user, and generate a checklists. Satisfaction expressions are activities that define actions the InterfaceAgent can perform. The activities involve:

- IdentifyUser, which include AuthenticateUser and CheckAccessRights.
- PassRequest.
- GenerateChecklist, which involve ProduceChecklist. InformUser

Security constraints defined are as follows:

- IF input, i.e., the user login information is incorrect THEN access shall be denied
- Login attempts shall be monitored and logged

To demonstrate message exchange between the agents in the multi-agent system, Unified Modeling Language (UML) is utilized. An example of message exchange between the agents is depicted in Figure 1.

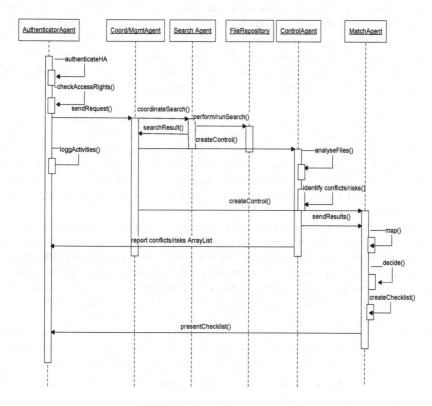

Fig. 1. Message exchange protocol

When AuthenticatorAgent receives a request from HA, the request is validated and if all constraints are satisfied, then request is passed to CoordinationAndManagementAgent. Coordination/Management agent creates control and match agents by cloning itself. Software agents verify links between documents and retrieve documents to map.

Control agent is responsible for analyzing goals and verification of requirements. Security requirements are derived from stakeholders' requirements, threats, and vulnerabilities. Meta agents analyze requirements to identify errors. In response to errors a problem report is generated and the process is halted until new request or task is received. If no errors are found the analyses result is send to match agent that provides mapping. The meta-agents provide mapping between security requirements and threats, as well as threats and countermeasures. Mapping to relevant expert knowledge stored in knowledge base performed in order to retrieve success/failure scenarios.

Meta-agents perform content analysis and mapping in order to create more complete solutions, while SoftwareAgents perform searching for documents and security ontologies.

The ontologies, for example ´Threat´ ontology and ´SecurityControl´ ontology, are retrieved from a local knowledge base [13]. The meta-agent uses a knowledge base, which contains facts and rules, to compare the ontologies, and an interpreter to match and execute rules. Rules can be constructed by terms or rules with several different alternatives, such as synonyms. The interpreter browses through the ontology with knowledge from the knowledge base. The ontologies are checked against the knowledge base with tags like *owl:Ontology*, *owl:Class*, *rdfs:subClassOf*, and *owl:onProperty*. Ontologies must contain related parts in order to make the mapping possible [13].

The search request can contain terminology, as well as classes and/or properties. The search and filtering the right documents are performed according to some criteria. The ManagementAgent merges the search result from the SoftwareAgents and passes it to the InterfaceAgent. Thus, the meta-agent perform mapping by comparing different ontologies. If there is a direct match the meta-agent can continue to work with the next part of the ontologies. Meta-agent is, also, able to combine ontologies in order to produce more complete result.

Security patterns contain documented knowledge of security professionals where a specific problem is addressed and solution described. Hence, security patterns can provide developers with the important information. Therefore, security patterns are included in the security ontology and are designed by using the ontology-based techniques. The ontology-based technique provides reusable and structured security information. Due to the OWL representation, the security patterns are available in a machine readable format and can be utilized in the multi-agent system [12].

Conclusively, multi-agent system can enable retrieval of knowledge from security ontologies and patterns and, hence, provide developers with the information about a specific threat or vulnerability, the impact of the threat on the particular security property, as well as information about possible solutions in order to minimize or prevent a particular security issue.

5   Conclusion and Further Work

In this paper, we have emphasized some crucial problems in development process in order to elucidate the essence of security issues. Security impacts core business processes in every organization. To prevent a number of cybercrimes and simple mistakes, security of e-commerce systems should be considered from the very beginning, i.e. early stage of the e-commerce software development. We have discussed the significance of the software security and necessity of traceability of security requirements.

To enable implementation of automated controls, support developers during software development and to provide visibility over the development process, multi-agent system that can fulfil these tasks, have been presented.

To search for patterns, to analyze the content and combine the ontologies in order to create more complete solutions according to the user request, agents in multi-agent system are applied. Moreover, ontology mapping performed by agents can save time and reduce human errors, which may help to increase security of the intended system. The multi-agent system, presented in this research, enables traceability of security requirements. The design of multi-agent system concerns with how agents cooperate to achieve goals, and what is required of each individual agent in order to accomplish the goals.

References

1. Cernuzzi, L., Juan, T., Sterling, L., Zambonelli, F.: The Gaia Methodology: Basic Concepts and Extensions 11, Part II, 69–88 (2004), doi:10.1007/1-4020-8058-1_6
2. Davis, N., Howard, M., Humphrey, W., McGraw, G., Redwine, S., Zibulski, G., Graettinger, C.: Processes to Produce Secure Software. In: Redwine Jr., S.T., Davis, N. (eds.) Software Process Subgroup of the Task Force on Security across the Software Development Lifecycle, vol. 1 (March 2004)
3. Fasli, M.: On agent technology for e-commerce: trust, security and legal issues. The Knowledge Engineering Review 22(1), 3–35 (2007)
4. Goertzel, M.K., Winograd, T.: Enhancing the Development Lifecycle to Produce Secure Software. A Reference Guidebook on Software Assurance, Technical Report, DACS (October 2008)
5. Jennings, N.R., Norman, T.J., Faratin, P., O'Brian, P., Odgers, B.: Autonomous Agents for Business Process Management, pp. 145–189. Taylor & Francis (2000) 0883-9514/00
6. Marik, V., McFarlane, D.: Industrial adoption of agent-based technologies. IEEE Intelligent Systems 20(1), 27–35 (2005),
   doi: http://dx.doi.org/10.1109/MIS.2005.11
7. McGraw, G.: Software Security Building Security. Addison-Wesley Pearson Ed. (2006) ISBN 0-321-35670-5
8. Moradian, E.: Secure transmission and processing of information in organisations systems. International Journal of Intelligent Defence Support Systems 2(1), 58–71 (2009)
9. Moradian, E., Håkansson, A., Andersson, J.-O.: Multi-Agent System Supporting Security Requirements Engineering. In: SERP 2010 - The 2010 International Conference on Software Engineering Research and Practice (WorldComp 2010), vol. 2, pp. 459–465. CSREA Press, USA (2010)
10. Moradian, E., Håkansson, A.: Controlling Security of Software Development with Multi-agent System. In: Setchi, R., Jordanov, I., Howlett, R.J., Jain, L.C. (eds.) KES 2010, Part IV. LNCS, vol. 6279, pp. 98–107. Springer, Heidelberg (2010)
11. Moradian, E., Håkansson, A.: Software Security Engineering Monitoring and Control. In: SAM 2011 The 2011 International Conference on Security and Management (WorldComp 2011). CSREA Press, USA (2011)
12. Moradian, E., Håkansson, A., Andersson, J.-O.: Security Patterns for Software Security Engineering. Accepted at the 16th International Conference, KES, San-Sebastian, Spain, September 10-12 (2012)
13. Moradian, E., Håkansson, A.: Ontology Design and Mapping for Building Secure E-Commerce Software. Accepted at the 8th International Conference on Web Information Systems and Technologies, Porto, Portugal, April 18-21 (2012)
14. Moradian, E.: Integrating Security in Software Engineering Process: The CSEP Methodology, KTH Royal Institute of Technology (2012)
15. Pfleeger, S.L.: Software Engineering Theory an Practice, 2nd edn. Prentice-Hall, Inc. (2001) ISBN 0-13-029049-1
16. Rice, D.: Geekonomics The Real Cost of Insecure Software. Pearson Ed. Inc. (2008) ISBN 0-321-47789-8
17. Van Vliet, H.: Software Engineering Principles and Practice, 2nd edn. John Wiley and Sons (2004) ISBN 0-471-97508-7
18. Van Gigch, J.P.: Applied General Systems Theory, 2nd edn. Harper & Row Publishers, New York (1978) ISBN 0-06-046776-2; Copyright 1978 by Van Gigch, J.P.
19. Wooldridge, M.J.: Introduction To Multi-Agent Systems. John Wiley and Sons Ltd. (2002) ISBN 9780471496915

Conceptual Ontology Intersection for Mapping and Alignment of Ontologies

Anne Håkansson and Dan Wu

Department of Software and Computer Systems
KTH Royal Institute of Technology
Forum 100
SE-164 40 Kista, Sweden
{annehak,dwu}@kth.se

Abstract. Combining ontologies can enrich knowledge within a domain and support the development and use of advanced services. This requires matching and combining the relevant ontologies for specific services, which can be supported by mapping and alignment of several ontologies. However, these techniques are not enough since the ontologies are often heterogeneous and difficult to combine. To overcome these problems, a conceptual ontology intersection is provided to map and align contents of the ontologies. This intersection is a conceptual ontology bridge between ontologies and contains parts from the involved ontologies. The contents are extracted by syntactic mapping and synonym alignment using an ontology repository, a rule base and a synonym lexicon using agents. The result is a set of concepts that together constitute the intersection, which is used for combining new incoming ontologies and, thereby, providing complex services.

1   Introduction

An ontology is a body of formal knowledge representation used in all kinds of domains. The ontology is a specification of a conceptualisation with objects, concepts and other entities and relations among them [1]. The ontology is used for enabling knowledge sharing and knowledge reuse, as well as, describing domain of discourse via systems that define a set of representational terms. The contents of the ontologies are human-defined texts describing the meaning of the names, the formal axioms that constrain the interpretation and the use of these terms. Hence, concepts of a domain is represented in a ontology as a model of the domain with which it is possible to reason about the entities within that domain.

Although, an ontology is a formal representation [2] that provides a semantic representation of a domain containing definitions of classes, relations and functions [3], the ontology cannot lend itself as an easy and direct solution to ontology mapping and integration. The problem is that using a well-formed description logic language for ontologies [4] is not enough since the semantic differences of ontologies are not reduced. The differences often depend on the ontology developers' perspectives and

A. Håkansson & R. Hartung (Eds.): *Agent & Multi-Agent Syst. in Distrib. Syst.*, SCI 462, pp. 105–124.
DOI: 10.1007/978-3-642-35208-9_6   © Springer-Verlag Berlin Heidelberg 2013

terminologies. The developers of ontologies most certainly have divergent perspectives of a domain and various purposes with the developed ontologies. Moreover, the terminology, chosen by the different developers, can be distinct. Therefore, when matching and combining the ontologies, syntactically and semantically, all these challenges must be encountered.

To support tasks envisaged by a distributed environment, one single ontology is not enough. Instead multiple ontologies need to be accessed from several applications [5]. There exist a lot of ontologies that taken together can enrich knowledge within a domain. Beside supporting knowledge sharing and reuse, the ontologies support the development and use of advanced services. This requires matching and combining the relevant ontologies for the specific services, which can be carried out by mapping and alignment of these ontologies. Mapping can provide a meta-level from which several ontologies can be accessed, corresponding concepts mapped and alignment applied to establish relations between equivalent the vocabularies of two ontologies and, thereby, increase the set of concepts in the intersection.

When several ontologies are involved in reasoning on the semantic web, as well as, combining ontologies to provide services on the distributed systems, the heterogeneousness of the ontologies becomes a problem. The ontology heterogeneities occur on different levels, such as syntactic, terminological, conceptual and semiotic levels [6] but also at semantic and pragmatic levels. Some syntactic matching solutions have been promising but the semantic mapping among the ontologies is still a difficult issue. Moreover, mapping on the syntactic level is not enough to obtain a useful combination of several ontologies for advanced services.

The research, presented in this chapter, tackles the heterogeneous problem by introducing a conceptual ontology intersection that serve as a bridge between involved ontologies. The extraction of the ontologies' contents is performed in several steps. A syntactic mapping is carried out by extracting the syntax of each ontology, comparing those by matching between the ontologies and storing the result in an intersection meta-model. Then, this intersection meta-model is used for handling semantics by synonym alignment. Synonyms are collected, aligned and stored as conceptual ontology intersections [7]. The result is intersections with concepts and synonyms that correspond to the parts that are found in both ontologies, which can be used for providing advanced services.

The chapter is structured as follows: Section two gives a brief description of constituents of ontologies and Section three is about methods for combining several ontologies. Section four describes a conceptual ontology intersection resulting from mapping and aligning several ontologies and violation handling. Section five presents related work and section six conclude this chapter.

2   Constituents of Ontologies

An ontology is defined as a set of representational primitives with which it is possible to model a domain of knowledge or discourse [7]. In other words, ontology is "a catalog of the types of things that is assumed to exist in a domain of interest D from the

perspective of a person who uses a language L for the purpose of talking about D.",
[8, 9]. Hence, the ontology states characteristics of a domain, which is known to be
true about that domain [10].

To effectively use ontologies, a well-designed and well-defined ontology language
should be used for development [10]. Some ontology languages, such as OWL 2, are
based on description logics, which is monotonic and adhere to the open world assumption [11]. With the open world assumption and monotonic reasoning, reasoning
can be used to derive implied relations in the ontology. With these assumptions, when
adding anything during the reasoning, it does not reduce the set of consequences.
Instead, the number of consequences will remain the same or increase. Applying the
reasoning on the contents of the original ontology without changing anything is an
important feature in the research of this paper since it ensures the quality of an ontology, which, for example, is avoiding non-contradictory concepts.

There is a set of constituents with which ontologies are built. These fundamental
building blocks are individuals (objects), attributes (properties), classes, relations,
function terms, axioms, restrictions, rules, and events [12]. The individuals are concrete objects in the domain and to describe these objects, they are related to attributes
that define properties with features or characteristics. These attributes, themselves,
can be objects or classes. Objects denote a concrete or abstract thing; class notation
describes individuals as collections of objects. The classes are concepts describing
abstract groups and sets of objects, which are defined by values of aspects constrained
as being members of a specific class [13].

To specify how objects are related, relations are applied. These relations can be a
of a particular type, or class, specifying in what sense objects are related. From relations, complex structures, called function terms, are formed to be used in place of an
individual term in a statement.

The ontology contains a set of axioms and facts in the domain [14], which together comprise the overall theory that an ontology describes. These axioms apply
constraints on sets of classes and the types of relations permitted between them. The
vocabulary can be defined with class axioms, and properties, such as ObjectProperty,
DataProperty, AnnotationProperty, Datatype, NamedIndividual and AnonymousIndi-
vidual. Class axioms allow establishing class relations, for example, the subClassof
axiom states that instances in one class expression are also instances of another class.
The classes can also express the semantically equivalences of the classes.

The ObjectProperty connects individuals in the ontology and the ObjectProperty-
Domain expresses that the individuals are from the domain, specified by the class
connected by that property. By using the object property axioms, domain and range
properties can be described. For example, the ObjectPropertyRange expresses that
individuals are from the range of the class connected by the property. This set of entities constitutes the signature of the ontology, which will be used for reasoning and
mapping.

The rules are production rules in Horn-clause form that describe the logical
inferences drawn from premises in a particular form [15]. Rules are useful for representing contingent features, such as, the relations between preconditions and postcon-
ditions [5]. They can capture a significant fragment of ontologies, by capturing simple

frame axioms and more expressive property axioms. The fragments permit, for example, stating that a class is a subclass of a class, which is useful for the research presented in this chapter. Drawing conclusions from the classes assure the quality of the mapping result.

Moreover, the ontologies contain the representation of entities, ideas, and events, along with their properties and relations. The events can be, for example, people, organizations, places, addresses, artifacts, phone numbers, and dates but can also be transactions. Also these parts can be used in the inference process to make the reasoning more reliable.

3   Methods for combining Contents of the Ontologies

Many methods have been developed for extracting and combining ontologies. Some of the most common are matching, mapping, alignment, merging, and integrating. Depending on the expected result of the extraction and/or combination, these methods can be beneficial.

Ontology *matching* aims at finding correspondences between related entities of different ontologies, syntactically and semantically [16]. The matching process takes ontologies as input, which consist of a set of entities like classes, properties, and determines correspondences as output, which are relations holding between the entities [17]. These correspondences can be equivalence, consequence, subsumption, or disjointness between ontology entities [16]. The matching can involve matching of the labels of the ontologies measuring the similarities, or distances, among nodes using matching estimations, which estimate the distance differences [18]. Unfortunately, there are different heterogeneities, such as syntactic heterogeneity, terminological heterogeneity, conceptual heterogeneity and semiotic heterogeneity, and even minor name differences and small structure variations can lead to matching problems.

Ontology *mapping* handles a part of the more advanced tasks concerning the alignment and merging of ontologies [5]. Ontology mapping is the task of relating the vocabulary of two ontologies sharing the same domain of discourse [5]. The mapping is mostly concerned with the representation of correspondences [19] where the structure of ontologies and their interpretations are specified as ontological axioms, which are preserved although mapping. Mapping one ontology to another ontology means that there is a corresponding concept, for each concept in one ontology in the other ontology that has the same or similar semantics [17]. The mapping can be partial since there might be concepts in a ontology that have no equivalents in the other ontology [20]. The mapping needs to map the contents of the ontologies regardless of the format of the concepts.

Ontology *alignment* is the task of establishing binary relations between the vocabularies of two ontologies [5] but does not depend on the choices of names in either ontology [20]. Ontology alignment is the task of creating links between two original ontologies and they are equivalent if a concept or relation in one ontology maps to a concept or relation in the other ontology. Ontology alignment is usually carried out if the sources are consistent with each other but are kept separate [21] and when they have complementary domains. Before two ontologies can be aligned, it may be necessary to introduce new subtypes or supertypes of concepts or relations in either one of the ontologies in order to provide suitable targets for alignment [20]. No other changes to the axioms, definitions, proofs, or computations in either ontology are made during the process of alignment. Alignment is the weakest form of integration since it requires minimal change [20]. Hence, it is useful for classification and information retrieval, but it does not support deep inferences [20].

Ontology *merging* is often used when the goal is to create a single coherent ontology that includes the information from all the sources and when the sources must be made consistent and coherent with one another but kept separately [21]. The merge is based on the discovery of the correspondences [19] by finding commonalities between two different ontologies and deriving a new ontology [20]. The new ontology may replace the merged ontologies but it can also be used as an intermediary. When performing ontology merging, a new ontology is created which is the union of the source ontologies, based on the correspondences between the ontologies [19]. The merged ontology captures all the knowledge from the original ontologies. The challenge in ontology merging is to ensure that all correspondences and differences between the ontologies are reflected in the merged ontology.

Ontology *integration* is the process of finding commonalities between two different ontologies and deriving a new ontology that facilitates the interoperability of the systems that are based on the integrated ontologies [22]. The term ontology integration has been used when the tasks are building new ontologies reusing other available ontologies, and integrating ontologies into applications but also merging ontologies into a single one that unifies them [24]. Integration is often used interchanged with ontology merging but a difference is that integration builds a new ontology whereas merging inserts one ontology into another ontology. The ontology integration is the process of building an ontology by assembling and extending other already existing ontologies, which becomes parts of the resulting ontology [24]. The new ontology may replace the other ontologies or become an intermediary part. There are three levels of integration: Alignment, mentioned above, Partial Compatibility and Total Compatibility (also called ontology merge). Partial Compatibility is an alignment of two ontologies that supports equivalent inferences and computations on all equivalent concepts and relations [25]. If two ontologies are partially compatible, any inference or computation, which can be expressed in one ontology by the aligned concepts and relations, can be translated to an equivalent inference or computation in the other ontology [25].

Since ontology mapping is the task of relating the vocabulary of two ontologies and concerned with the representation of correspondences specified as ontological axioms, mapping is used for finding the similar concepts. The corresponding concepts that have the same or similar syntax and semantics are recorded for further mapping tasks. When it comes to context for the ontologies, ontology alignment is applied because it is used to create links between two original ontologies, i.e., if they are equivalent. The ontologies are expected to be consistent and be in complementary domains but will be kept separately. To align the ontologies, new synonyms are introduced to provide ontologies as suitable targets for alignment.

4   The Conceptual Ontology Intersection

The conceptual ontology intersection, presented in this chapter, builds on the contents of the ontologies. The contents are extracted by syntactic mapping including syntactic matching, to build a intersection of all syntactic related parts, and synonym alignment to enrich the intersection with synonyms. This intersection becomes a conceptual ontology bridge between ontologies and contains related parts found in the involved ontologies.

A process for conceptual ontology intersection is used for mapping and alignment, see Figure 1. The process starts with fetching the ontologies from the web. Then, the syntactic mapping, with syntactic matching takes place. The result is a meta-model with syntactic parts found in the ontologies. Next step is a synonym alignment, which takes either ontologies or meta-model as input and expands the meta-model with more parts, this time with synonyms. Since mapping and alignment can introduce violations in the meta-model, violations are checked.

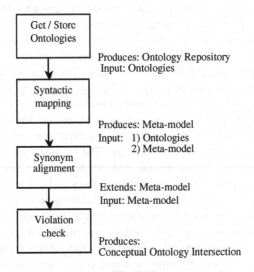

Fig. 1. The process

For each time a new ontology is introduced the meta-model will expand and the violation check needs to go through the whole meta-model.

4.1   The Syntactic Mapping

Mapping the ontologies syntactically is to match and map each concept in one ontology to each concept in the other ontology. The matching is carried out by picking up the different tags and words in the ontologies, using agents. The tags are the standard xml-tags that are used for, e.g., URL, texts, pictures, and signatures [26]; whereas the words are concepts in a domain. The mapping is to reason with the contents of the ontologies to find relations and correspondances between the concepts.

To match and map parts in the ontologies, a rule base is used. The rule base includes parts that can be found in the ontologies. Some commonly used ontology parts are basic metadata, document metadata, RDF metadata, and description term:

- The basic metadata includes *url:, desc:, def:, ref:, pop:,* and *ns:,*
- The document metadata consists of *hasEncoding:, hasLenght:, hasMd5sum:, hasFiletype:, hasDateLastmodified:,* and *hasDataCache.*
- The RDF metadata has *hasGrammar:, hasCntTriple:, hasOntoRatio:, hasCntSwtDef:* and *hasCntInstance.*
- For the description term, there are other terms associated to it. For example, *log:forSome, title, s:label,* and *s:subClassOf.* To the term "Class", there are terms like *s:comment, s:label, ont:UniqueProperty, s:domain,* and *s:range.*

The terms are represented as facts and/or rules, in the rule base, to be able to compare the contents of the ontologies. In the rule base, there are single valued facts and compound terms, which are rules with different levels of complexity. Some terms can be matched directly to the facts whereas other terms must use rules to reach a conclusion, i.e., use reasoning to extract the contents of the ontology. By implementing all these terms in a rule base, the rule base will contain necessary knowledge and guidance to perform more efficiently.

Direct matching or *syntactic matching* is to match every line in ontology to the other ontology, which is to check that the tags and words are the same in both ontologies. Also words with different tags are matched between the ontologies. The process of direct matching is as follows: The process starts with one ontology and extracts the first tag and match it with the other ontology. Then, the process use the word, attached to the tag, and match it to the other ontology tags' word. For each match, the tags together with the words are stored in a meta-model as pair <Tag, Word>.

Then, the matching process proceeds with only the words, ignoring the tags, and matches each word in the ontologies. If the word appears in both ontologies, irrespectively of the tags, the tags and words are stored in the meta-model as <Tag1, Word>, <Tag2, Word>. The result from direct matching is a meta-model with same tags and words, or different tags and words, found in the ontologies. However, to know whether the words have correct connections to each other, reasoning with the contents is carried out. Reasoning with the ontologies capture context, to some extent, and might prevent that the mapping leads to wrong result.

For *syntactic mapping* with reasoning, the result, from direct matching in the meta-model, is used to reason with other ontologies, which is performed by using rules from the rule base and the ontology repository. These rules are parsing rules, which for each pair, stored in the meta-model, capture the surrounding environments in the ontologies by comparing the surrounding terms connected to the pairs, i.e., tags and words. The syntactic mapping use words in one ontology and rule base and, then, map with an other ontology to extracts concepts and relations with connections between classes, objects, and properties.

The rules for basic metadata, document metadata and RDF metadata, with axiom and entities, and description term are used to handle the syntactic contents of the ontology. The same types of entities and their contents are syntactically compared. For the types, the labels of the class are compared between the ontologies and, for the contents, the labels of the object property are compared. For example, a class have a subclass, and they share a property relation.

The syntactic mapping compares the concepts and relations in one ontology with the contents of another ontology by following strategies:

1) the letter cases are ignored. No matter the letters used in a word, it is consider to be identical in both ontologies, regardless the combination of the upper cases and lower cases.
2) only the letters are compared, and special characters are excluded.
3) grammatical forms are ignored, i.e., singular and plural of nouns are equal and all the forms of verbs are ignored.
4) nothing is excluded. As long as the definitions are not in conflict with each other, they can coexist and enrich the knowledge but only the same concepts and relations are checked and stored.
5) non-matched parts in the ontologies are left without consideration. These parts can still be reached by looking in the original ontologies.

The result from mapping is all the entities and the properties found in the ontologies that are comparable. The entities are stored as concepts in the meta-model and properties are stored as relations. Moreover, to keep the concepts and related relations connected to each other, these are connected by signatures, which are stored as rules, in the meta-model. Also, the relations are stored as rules in the meta-model.

To illustrate the syntactic matching and mapping with reasoning, an example is given below, see Figure 2. In the figure, classes, subclasses, objects and properties are presented. In Ontology 1, left hand in Figure 1, Document is a subclass of Root. The subclass connects to other parts by properties, which, in this case, are: "date-creation", "name", "has-author", and "has-topic". These properties connect the Document to several objects, which are Literal, Author, and Topic. Other connections, i.e., Journal, Publication, Book, Presentation, and Report are subclasses to Document. Ontology 2, on the right hand of the figure, shows that Document is a subclass of Source and contains the subclasses Website, Publication and Ontology. To the Document class, also the property "has-author" is connected, which is another way of connecting Document and Source.

The meta-model is expanded for every new axiom (fact) and entity (concept) and property (relation) the reasoner finds in the ontology. Also, the set of signatures, that is considered to be as being equivalent by the syntactic matching and syntactic mapping, are stored in the meta-model. A partial result of reasoning with Ontology 1 and Ontology 2, for the example above, is O1={(root, class), (document, subclass), (publication, subclass), (has_author, objectproperty)}, O2={(source, class),

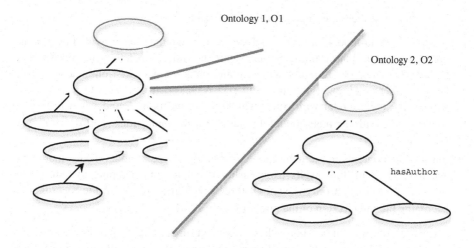

Fig. 2. The example ontologies

(document, subclass), (hasAuthor, objectproperty)}. These sets are stored as rules in the meta-model, as well as, the relations between them. The mapping produces following rules:

Class(X):- 'Root'.
Class(X):- 'Source'.
Class(X):- 'Publication'.
Class(X):- 'Paper'.
Subclass(X):- 'Document'.
Subclass(X):- 'Publication'.
Object(X):- 'Author'.
ObjectProperty(X):- has-author.
ObjectProperty(X):- hasAuthor.
SubClassOf(X, Y):- Subclass(X), Class(Y).
ObjectPropertyRange:- Subclass(X), ObjectProperty(X), Object(X).
ObjectPropertyRange:- Subclass(X), ObjectProperty(X), Class(X).

These rules expand the meta-model. One can argue that the rule base should be expanded for these new developed rules for matching and mapping. The argument for separating the rules for matching/mapping from rules for tags is that the rule base will be used for new comparing activities and the meta-model will be built from the beginning for each new comparison.

If a third ontology is compared with the other ontologies, the meta-model, and rule-base are used. However, since the third ontology might have more connections that can found in either of the already matched and mapped ontologies, also the ontology repository needs to be involved.

From this point there are three approaches provided:

1) expand the meta-model with rules, which are new connections provided by matching and mapping both ontologies but one at the time, Expanding the meta-model with new connections requires matching and mapping with first one ontology and then the other ontology giving the result {O1 ∧ O2, O1 ∧ O3, O2 ∧ O3}
2) expand the meta-model with rules, which are connections found when matching and mapping with the meta-model. Expanding with connections is {O1 ∩ O2, O1 ∩ O3, O2 ∩ O3}
3) limiting the contents in the meta-model. Limiting the meta-model means matching and mapping with the results from earlier matching and mapping to reduce the rules so the rules that only covers all three ontologies are left in the meta-model becoming the intersection of the ontologies, i.e., {O1 ∩ O2 ∩ O3}.

By using any of these three approaches, this meta-model becomes a bridge between the ontologies, containing parts found in all ontologies. However, all three approaches have advantages and disadvantages, and these depend on the situation in which the matching and mapping is performed. The first approach will include everything in the ontologies, i.e., not exclude anything, while the other two approaches are intersections of the ontologies with different sets as shown above. In the research in this chapter, the third approach is employed.

4.2   The Ontology Repository

The ontology repository is designed to store the ontologies. The ontologies are stored in their original form and the meta-model is used to sort the repository and retrieve the ontologies. Since the meta-model describes the ontologies, it forms context for the ontologies. The context also contains information about the ontology's annotation, i.e., the purpose, specific date, particular interest and the owner. The purpose of an ontology is the intention or goal given by the owner or developer of the ontology; the interest is the users, groups or organizations for which the ontology is developed. The owner of the ontology is the person or the organization that owns the copyright of the ontology. The date is the date of developing the ontology.

Consider the example ontologies in Figure 2, the Ontology 1 is about researchers and Ontology 2 is an ontology for inferring on the web. To store those ontologies in the repository, following parts can be used see Table 1.

Table 1. Example of the annotation

	Purpose	*Owner*	*Interest*	*Date*
Ontology1	Application	Karlsruhe		071010
Ontology2	Online course	Stanford	Graduate credits	100920

Using annotation can help to enrich the meaning of the integration. This can be applied to the meta-model and give more information about the integration.

4.3   The Synonym Mapping Process - Synonym Alignment

Synonym alignment is the process of finding synonyms for the words in the ontologies and checking whether or nor they match with the contents in the other ontologies. There are two slightly different processes for the alignment, which depend on the number of ontologies to align: 1) two ontologies are aligned and 2) three and more ontologies are aligned. The first process fetch synonyms from the contents of the ontologies and the second from the meta-model and the ontology to be aligned.

If two ontologies are aligned, the meta-model contains only the concepts that have syntactic match but there might be concepts that are similar in the original ontologies. Therefore, the whole contents of the two ontologies must be used for synonym alignment.

If three or more ontologies are aligned, it is enough to use the meta-model, instead of the two original ontologies because, this time, the meta-model contains both the syntactic match and synonym matching. This meta-model is used when the third or more ontology is being aligned.

When ontologies are aligned, the parts: classes, objects, and properties, in the ontologies, are used to find synonyms. For each of these parts, synonyms are searched and fetched from the online WordNet [14] giving back a set of words in the form {< , >,...,< , >}. The set of synonyms is used to check if the classes, objects and properties belong to the same set of synonyms and, if so, they are aligned. The alignment between the synonyms is creating links between two original ontologies. They synonyms are equivalent if a synonym of a concept or relation in one ontology map to 1) a concept or relation in the other ontology, or 2) a synonym of a concept or relation in the other ontology. If the synonyms are equivalent, they are stored in the meta-model, as the <concept, synonyms> or <relation, synonyms>. If synonyms are not equivalent, the synonyms are discharged.

When three or more ontologies are aligned, the process of finding and aligning the synonyms is identical but instead of using the original ontologies, that are already aligned, the meta-model with syntactic mapping and semantic alignment is used together with the third or more ontologies. The concepts and the relations in meta-model as well as the contents, i.e., classes, objects and properties, of the third or more ontologies are used to find the synonyms. Only the contents of the third or more ontologies are sent to WordNet [14] to find synonyms. The set of synonyms are aligned with the contents of the meta-model and if synonyms are equivalent, these synonyms are stored in the meta-model.

Commonly, words used in the ontologies are nouns and verbs. However, WordNet [14] generates both nouns and verbs as synonyms. That is, if a noun (or a verb) is sent to WordNet, it generats both nouns and verbs synonyms. This must be limited so noun give nouns as synonyms and verb gives verbs as synonyms. For synonyms to ontologies this means, the class and objects commonly are nouns, and hence, the synonyms, fetched from WordNet, must be a noun; the properties can be verbs and, therefore, the synonyms must be verbs. Limiting the synonyms to word classes can make the alignemnt more correct since, other word classes, might introduced missinterpretation. For example, the noun "document" and the verb "document" give different sets of synonyms.

To illustrate the synonym alignment, an example of synonyms for the Ontology 1 and Ontology 2 mentioned above, is given. O1 includes the word "Document" and when looking up synonyms in WordNet [14], it provides the synonyms: "written document", "papers", "text file" together with explanations, such as "writing that provides information (especially information of a official nature)", a "written account of ownership or obligation" and "(computer science) a computer file that contains text (and possibly formatting instructions) using seven-bit ASCII characters". It also provides synonyms for the verb "document" but only with the explanations: "record in detail", "support or supply with references". Since Document is a noun, the nouns are in focus when comparing (document, class) and (paper, class). Since WordNet [14] gives "paper" as a synonym to "Document", it is found to be an alignment between those words – even though the ontologies use two different words. Both these words are stored in the meta-model.

4.4   Violation Handling for Syntactic Mapping and Synonym Alignment

Although syntactic mapping and synonym alignment are made, the contents of the ontologies may not harmonize. To find contradictions or different implications among the parts in the ontologies, violation check for the syntactic mapping and synonym alignment is carried out. The concepts together with the relations are checked for violations to find how well they really match, map but also align.

Violation Check for Syntactic Mapping

For syntactic mapping, including syntactic matching, the combination of classes, objects and properties can mismatch between the meta-model and ontologies. The violation check for syntactic mapping controls if the concepts and relations really share corresponding concepts and relations between the mapped ontologies. The violation check also controls if relations are connecting the same concepts in the ontologies. The checking is both carried out by syntactic mapping, which controls if the concepts and relations are the same, as mentioned above, by following the five different strategies, such as, accepting different ways of writing the words but also the implications, i.e., O1A -> O1B $\wedge$ O2A -> O2B.

To illustrate contradictions, an example of the ontologies, O1 and O2 presented above, is given. The result of a syntactic mapping is that the relations "has-author" and "hasAuthor" are equal and the concepts Document, Author and Source are stored. In the Ontology 1, the subclass Document is connected to the class Author and in the Ontology 2, the subclass Document is connected to class Source. Hence, there is a contradiction between the classes Author and Source.

To handle the contradiction, rules from the rule base are used to solve the problem. The rules have three different outcomes and the possible results are checking the kind of violation and the outcomes are: 1) accept violation, 2) provide notification, or 3) confirm violation. Hence, the kind of violation found between the ontologies steers the outcome.

The first type of violation, accept violation, implies that relations are not violating anything between the ontologies, even though the concepts are the same same type of entity. This kind of violation implies that the relations in the ontologies are found and they are equal but the names of the classes or objects differ. The result is stored as an equivalent violation case in the conceptual intersection ontology, i.e., in the given example the subclass has the classes Source and Root, together with the related concepts, and relations. The format for the case is EquivalentViolation(Document, < Root, Source >) and the reason is that the synonym violation needs to check this equivalent violation case.

The second kind of violation, provide notification, also implies that relations are not violating anything between the ontologies, but, in this case, the concepts are of different same types of entities. This violation implies that the relations in the ontologies are equal but the types of entities differ. The result of this violation is stored as a notification in the conceptual intersection ontology with the concepts, relations, and the different types. In the example mentioned above, one entity is a class and the other entity is a subclass. The format for notifications is the NonEquivalentViolation(Document, <Publication, class $\wedge$ Publication, subclass>) where Publication is a subclass in O1 and a class in O2.

The third kind of violation, confirm violation, implies that the concepts and the classes violate too much to be stored in the conceptual intersection ontology. An example is that a concept and a relation have the same name. This case should not occur since the matching and mapping should avoid finding equivalence between concepts and relations from the tags and words. Nonetheless, this confirm violation case also comes into effect if none of the above cases can capture a violation. The confirmation of the violation is stored in the conceptual intersection ontology as ConfirmedViolation(Document, <HasAuthor, property $\wedge$ Author, class>)

The violation check process work through all of the concepts and relations in the meta-model and carries on until the whole meta-model is checked. The result with concepts, and relations that are acceptable and non-violating any of the syntactic mapping are stored in the conceptual ontology intersection.

Violation Check for Synonym Alignment

Also the synonyms are checked for violation problems. This is to check that the synonyms are comparable to concepts, and relations in the ontologies, and/or the meta-model. The violation check for synonyms is more difficult since it is difficult to detect synonym violations.

A violation check for synonyms is when syntactic mapping has found a violation, i.e., accept violation, where the relations in the ontologies are found to be equal but the names of the classes, or objects, differ. The synonym violation check uses the EquivalentViolation in the meta-model to look up the concepts in the WordNet and receive a set of the synonyms for these concepts. Then, the concepts are tested towards the set of synonyms to find equivalences. If succeeded, the result is stored in the conceptual ontology intersection as EquivalentCheck(concept, <concept, synonyms>) with the concepts and relations. In the case of the example, above, it can be

EquivalentViolation(Document, <Root, (beginning, origin, root, rootage, source) >, < Source, (beginning, origin, root, rootage) >).

Another violation that is when the meanings of the words differ. Meaning check requires semantic examination of the words to investigate the relations between the words. The violations can arise when a term, which is a concept or relation, in one ontology, *does not* align the same term, in the other ontology, even though surrounding concepts and relations *do* align. This means that almost the whole set of synonyms for one ontology, O1, is the same set of synonyms for other ontology, O2 and the aligned set is moved to the conceptual ontology intersection. The synonyms that do not align needs to be denoted and stored in conceptual ontology intersection by applying a notification to the synonym in the form of NonAlignedViolation(concept, synonym).

This violations can also arise when a term, , i.e., concept or relation, in one ontology, *does* align the same term, in the other ontology, but the surrounding concepts and relations *do not* align. Then, almost the whole set of synonyms for one ontology may be totally different from the other ontology's set. The synonym mapping can find a concept, or relation, in the involved ontologies, because it distinguishes itself from the set of the aligned concepts and relations, by using the surrounding concepts and relations to provide information about the domain and context. Again, the synonyms, that do not align, needs to be denoted and stored in conceptual ontology intersection by applying a notification to the synonym in the same form of earlier NonAlignedViolation(concept, synonym) together with the concept and/or relation that does align.

As result of the violation check or syntactic mapping and synonym alignment, a conceptual ontology intersection is produced with parts that are captured in the violation check.

4.5   Context of the Ontologies

As a result from mapping and alignment, the conceptual ontology intersection becomes a context for all compared ontologies. One entity is used to build context and a domain for that entity. All the ontologies that are using a specific entity can be fetched and incorporated in the system and in the intersection and, therefore, extend and strengthen the context for the entity. To use the context from different perspective, a perspective property is introduced as a relation in the ontology intersection. The perspective property is used to combine the parts of the ontologies that concerns the property. The perspective property is illustrated in Figure 3, where the conceptual ontology intersection is in the middle and the ontologies in the system is connected to the intersection by the relation "hasPerspective".

The contents in the ontologies can, together, give knowledge about a context in a domain. This useful facililty can provide information about, for example, a service and can, when combined with several different ontologies or similar intersections, provide complex services to the users. This requires that one ontology provides a service that can be combined with the services provided by the other ontologies.

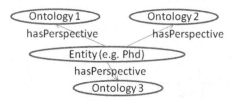

Fig. 3. Perspectives for the entities

For the context comparison, rules are used. The rules must prove that the entities of the ontologies match. The rules investigate the entities by checking that the contents of these entities belong to the same context. Thus, the rules check whether the contents of the entities are found in that domain or not, which is the context for the ontology. Conclusively, every ontology's entities are used to build the services.

Moreover, the rules can use other rules and facts to match the contents. If some information piece is missing, the system can turn to other sources, like the web or users to get additional information.

4.6   The Architecture

The architecture for the system distinguishes the ontology repository, rule base, meta-model and conceptual ontology intersection, see Figure 4. The ontologies, fecthed either from a file or an URL, are deposited into the ontology repository and the rules for matching, mapping, and alignment are stored in the rule base. These rules use, with the help of a rule engine, the ontologies to generate result for syntactic mapping. Syntactic mapping with matching applied the strategies 1) - 5) mentioned in 4.1 and the result is concepts and relations, which are stored in the meta-model.

The ontology repository is sorted accordingly to the contents of the ontolgies and with the help of the meta-model. The ontology repository stores signatures from the ontologies by adding the signatures as records in the table. These records are connections to the ontologies and are used for searching for the ontologies used in the system.

Synonym alignment provides function for synonym alignment, which uses the result in the meta-model to fetch synonyms in synonym lexicon and align the ontologies. The result from the synonym alignment is also stored in the meta-model.

The meta-model is checked for violations, syntactic and synonyms, and the concepts, relations and synonyms that passes the violation check are stored in the conceptual ontology intersection, as well as, the parts that are not contractions or wrong impliations. The conceptual ontology intersection constitiues the context for the ontologies mapped and aligned in the system.

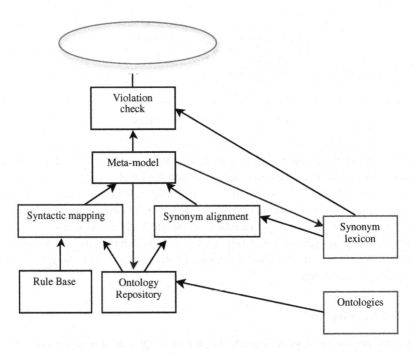

Fig. 4. Architecture

5   Related Work

A lot of semi-automatic and automatic methods have been developed for ontology ontology merging and mapping [27]. For example, FCA-Merge is a method for ontology merge [28] and IF-Map is a method for ontology mapping [29]. In the FCA-Merge method, natural language techniques are used to derive lattices of concepts. The lattices are explored, manually, by knowledge engineers who also build the merged ontology with the help of semi-automatic guidance of the FCA-Merge method [27]. Exploring the lattices and build merged ontologies can be a time-consuming and tedious job, if the ontologies are medium and larged sized.

IF-Map method is an automatic method for ontology mapping. IF-Map provides automated support in the alignment of ontologies by automatically generating mappings between a reference and various local ontologies. The method is a concept-to-concept and relation-to-relation mapping for local ontologies, which limits the result of the mapping.

Another automatic approach for mapping two ontologies is a process using weighted similarities [30]. The result is a set of statements that contains the semantic correspondences between similar elements in two ontologies with associated relations and a confidence of the mapping result. Confidences in results of mapping is an important feature and by applying rules for syntactic mapping, confidence can be upheld on a certified enough level [7].

Methods have been developed for ontology alignment. An example of ontology alignment is a general framework for alignment. This framework is data interlinking, which separate data interlinking and ontology matching activities to enhance data linking through ontology alignment by allowing linking specifications to reuse ontology alignments in a natural way [31]. Using specifications are interesting but should be extended together with the contents and relations found in the ontologies.

Another example of ontology alignment is an algorithm that uses a complete prover to decide subsumptions or equivalences between classes given initial equivalence of some classes and analysis of the relationships in the taxonomy [32, 33]. The complete prover, in its basic form, consists of "testing each possible variables assignment. A backtrack is executed when an inconsistency is reached where certain rules can be used in order to reduce the search space " [33].

Finding the equivalences and subsumptions is the main task for the research in this chapter but not from any initial equivalence between the ontologies, rather find all equivalences between the ontologies.

Another interesting method is semantic translations of the ontologies handling similar domains [34]. This method includes developing bridges for axioms in the ontologies to merge these ontologies. Ideas from this technique are used to build intersections of concepts and relations which become bridges between the ontologies. In the research, presented in this chapter, also synomyms are utilsed to expand the possibly to capture more bridges between the ontologies, that is when the ontologies use different vocabulary in the same domain.

6   Conclusions and Further Work

The main contribution of this chapter is conceptual ontology intersection which includes parts that are related in several different ontologies. Contents are extracted and compared by syntactic matching and syntactic mapping and synonym alignment. The syntactic mapping and synonym alignment are applied on ontologies fetched from the Web. These ontologies are stored in their original form in an ontology repository to be able to revisit and reuse them.

Contents are extracted by syntactic matching and mapping using a rule base, and the ontology repository. The rules are used for the matching and mapping process where the concepts and relations are extracted from the ontologies. During the mapping process, a meta-model of the ontologies is built. The meta-model contains the parts from the ontologies that have syntactically mapped and synonym matched. The meta-model is a conceptual ontology intersection, which is an ontology bridge containing syntactic similarities between the ontologies.

To enrich the connections between the ontologies, synonym alignment is used. The alignment is either aligning two ontologies or the result from syntactic matching and mapping ontologies and an ontology. To find the synonyms, the alignment uses a synonym lexicon and the result expands the meta-model with synonyms.

The result is a conceptual ontology intersection that contains parts from the ontologies. The ontology intersection provides contexts for the ontologies. Building contexts for the ontologies can help to reuse the ontologies but also for further mapping and alignment.

The solution works for small and medium sized ontologies. For large sized ontologies, links to the ontologies' sites are stored instead. Also, the synonym lexicon is fetched from the Web. The lexicon is used for searching and finding synomyms for alignment. The lexicon itself will not be stored in the system but the synomyms, that is results from the alignment, are stored. The rule base stores rules for mapping and alignment in the system.

References

1. Duc, T.T., Haase, P., Motik, B., Grau, B.C., Horrocks, I.: Metalevel Information in Ontology-Based Applications. In: Proceedings of the 23rd National Conference on Artificial Intelligence (AAAI 2008), Chicago, Illinois (2008)
2. W3C: OWL 2 Web Ontology Language, Structural Specification and Functional-Style Syntax, W3C Recommendation (October 2009)
3. Gruber, T.R.: Toward Principles for the Design of Ontologies Used for Knowledge Sharing, Technical Report KSL 93-04, Knowledge Systems Laboratory, Stanford University (1993)
4. Knorr, M., Alferes, J.J., Hitzler, P.: Local closed world reasoning with description logics under the well-founded semantics. Artificial Intelligence (2011)
5. Kalfoglou, Y., Schorlemmer, M.: Ontology Mapping: The State of the Art. Journal The Knowledge Engineering Review 18(1) (2003)
6. Bouquet, P., Ehrig, M., Euzenat, J., Franconi, E., Hitzler, P., Krötzsch, M., Serafinin, L., Stamou, G., Sure, Y., Tessaris, S.: D.2.2.1 Specification of a common framework for characterizing alignment, Knowledgeweb, realizing the semantic web (2005)
7. Wu, D., Håkansson, A.: Accepted in Proceedings of 8th International Conference on Web Information Systems and Technologies (WEBIST), April 18-21 (2012)
8. Gruber, T.: Ontology. In: Liu, L., Tamer Özsu, M. (eds.) Encyclopedia of Database Systems. Springer (2009),
   http://tomgruber.org/writing/ontology-definition-2007.html
9. Sowa, J.F.: Ontology (2011), http://www.jfsowa.com/ontology/index.html
10. Sowa, J.F.: Ontology, Metadata, and Semiotics. In: Ganter, B., Mineau, G.W. (eds.) ICCS 2000. LNCS, vol. 1867, pp. 55–81. Springer, Heidelberg (2000)
11. McKeon, M.: Model-Theoretic Conceptions of Logical Consequence. Internect Encyclopedia of Philosopy. Michigan State University (2004),
   http://www.iep.utm.edu/logcon-m/ (Originally published: June 16, 2004)
12. Knorr, M., Alferes, J.J., Hitzler, P.: Local closed world reasoning with description logics under the well-founded semantics. Artificial Intelligence (2011),
   doi:10.1016/j.artint. 2011.01.007
13. Gruber, T.R.: Ontolingua: A mechanism to support portable ontologies. Stanford University, Knowledge Systems Laboratory, Technical Report KSL-91-66 (March 1992)
14. Carlsson, C., Fullér, R.: Possibility for Decision. STUDFUZZ, vol. 270. Springer, Heidelberg (2011) ISBN 978-3-642-22641-0
15. Princeton University, WordNet, A lexical database for English,
   http://wordnet.princeton.edu/wordnet/ (last update: June 21, 2011)

16. Grosof, B.N., Horrocks, I., Volz, R., Decker, S.: Description Logic Programs: Combining Logic Programs with Description Logic. In: WWW 2003, May 20-24 (2003)
17. Jérôme, E., Pavel, S.: Ontology Matching. Springer, Heidelberg (2007) ISBN 3-540-49611-4
18. Ehrig, M., Sure, Y.: Ontology Mapping - An Integrated Approach. In: Bussler, C.J., Davies, J., Fensel, D., Studer, R. (eds.) ESWS 2004. LNCS, vol. 3053, pp. 76–91. Springer, Heidelberg (2004)
19. Kotis, K., Vouros, G.A.: The HCONE Approach to Ontology Merging. In: Bussler, C.J., Davies, J., Fensel, D., Studer, R. (eds.) ESWS 2004. LNCS, vol. 3053, pp. 137–151. Springer, Heidelberg (2004)
20. de Bruijn, J., Ehrig, M., Feier, C.: Ontology mediation, merging and aligning. In: Semantic Web Technologies Trends and Research in Ontologybased Systems, pp. 1–20 (2006)
21. Sowa, J.F.: Top-Level Ontological Categories'. International Journal of Human Computer Studies 43(5/6), 669–685 (1995)
22. Noy, N., Musen, M.: PROMPT: Algorithm and Tool for Automated Ontology Merging and Alignment. In: Proceedings of the National Conference on Artificial Intelligence, AAAI (2000)
23. Sowa, J.F.: Knowledge Representation: Logical, Philosophical and Computational Foundations. Brooks/Cole (2000)
24. Pinto, H.S., Gómez-Pérez, A., Martins, J.P.: Some Issues on Ontology Integration. In: Proc. of IJCAI 1999's Workshop on Ontologies and Problem Solving Methods: Lessons Learned and Future Trends (1999)
25. Pinto, H.S., Gomez-Perez, A., Martins, J.P.: Some Issues on Ontology Integration. In: Proc. of IJCAI 1999's Workshop on Ontologies and Problem Solving Methods: Lessons Learned and Future Trends (1999)
26. Sowa, J.F.: Principles of ontology, http://www-ksl.stanford.edu/onto-std/mailarchive/0136.html
27. Håkansson, A., Hartung, R., Moradian, E., Wu, D.: Comparing Ontologies Using Multi-agent System and Knowledge Base. In: Setchi, R., Jordanov, I., Howlett, R.J., Jain, L.C. (eds.) KES 2010, Part IV. LNCS (LNAI), vol. 6279, pp. 124–134. Springer, Heidelberg (2010)
28. Kalfoglou, Y., Schorlemmer, M.: Ontology mapping: the state of the art. The Knowledge Engineering Review 18(1), 1–31 (2003)
29. Stumme, G., Maedche, A.: Ontology Merging for Federated Ontologies on the Semantic Web. In: Proceedings of the International Workshop for Foundations of Models for Information Integration (FMII 2001), Viterbo, Italy (September 2001)
30. Kalfoglou, Y., Schorlemmer, M.: Information-Flow-Based Ontology Mapping. In: Meersman, R., Tari, Z. (eds.) CoopIS/DOA/ODBASE 2002. LNCS, vol. 2519, pp. 1132–1151. Springer, Heidelberg (2002)
31. Scharffe, F., Euzenat, J.: Linked Data Meets Ontology Matching - Enhancing Data Linking through Ontology Alignments. In: Filipe, J., Dietz, J.L.G. (eds.) KEOD 2011 - Proceedings of the International Conference on Knowledge Engineering and Ontology Development, Paris, France, October 26-29, pp. 279–284 (2011)
32. Dou, D., McDermott, D., Qi, P.: Ontology Translation on the Semantic Web. In: Spaccapietra, S., Bertino, E., Jajodia, S., King, R., McLeod, D., Orlowska, M.E., Strous, L. (eds.) Journal on Data Semantics II. LNCS, vol. 3360, pp. 35–57. Springer, Heidelberg (2005)

33. Euzenat, J., Valtchev, P.: Similarity-based ontology alignment in OWL-Lite. In: de Manta-ras, R.L., Saitta, L. (eds.) Proc. 16th of European Conference on Artificial Intelligence (ECAI), Valencia, ES, pp. 333–337 (2004)
34. Giunchiglia, F., Shvaiko, P.: Semantic matching. In: Proc. IJCAI 2003 Workshop on On-tologies and Distributed Systems, Acapulco, MX, pp. 139–146 (2003)
35. Mao, M., Peng, Y., Spring, M.: An adaptive ontology mapping approach with neural net-work based constraint satisfaction. Web Semantics: Science, Services and Agents on the World Wide Web 8, 14–25 (2009)

Implications and Solution
for High-Speed Business Architecture

Ronald L. Hartung

Franklin University
Columbus Ohio 43224
USA

Abstract. This chapter presents a forward looking view of the transition to extremely rapid business operations. The argument for this future is based on the
trajectory of business operations. It is not certain, but does show an interesting
future view for systems the IT professionals will have to build and manage.

1 Introduction

The term digital economy can encompass a range of applications and concepts. In our
reflection on this term, the concept of the electronic second self [5] comes to the fore.
Just as a part of our existence in now contained in an electronic world, so is commerce moving into the electronic world. This is not really a new phenomenon. One of
the first parts of commerce to move into an electronic domain was money itself.
Western Union was transferring money by telegraph and Teletype for a long time.
Entertainment became electronic with radio and then TV. Now the freedom of on
demand streaming is here. The web has brought us the ability to purchase in an electronic store. Actions are now part of the landscape (eBay, etc.).

So where does this lead next? Our proposal is the multi agent negotiation for a set
of interrelated commodities is a possible valuable business process. The makes the
purchase and delivery of supplies, the scheduling of production and the coordination
of customer orders and delivery a single unified process. This is a futuristic view and
may not be acceptable to business at this stage. This chapter presents an argument that
this is the likely future of business.

In a very real sense, this is the extension of "just in time" concepts that have been
accepted practice for many years. The inventory of parts and products is a business
cost. Just-in-time techniques were seen as a way to reduce these costs. If materials for
production were delivered as needed, a warehouse is not required, or at the least reduced in size. Even more, it is possible to keep the materials on the supplier's books
until they are actually needed, reducing inventory and reducing the capital required to
own the inventory, also reducing the taxes on the inventory. This is an example of the
acceleration of business that is favored by reducing cost.

This paper explores a possible future state for business. This is based on the three
aspects. The continued acceleration, introduced above is the main driving aspect. The
distribution of manufacturing and production into a worldwide network is already in
place. These two aspects are arguably well established.

A. Håkansson & R. Hartung (Eds.): *Agent & Multi-Agent Syst. in Distrib. Syst.*, SCI 462, pp. 125–136.
DOI: 10.1007/978-3-642-35208-9_7   © Springer-Verlag Berlin Heidelberg 2013

And third, the opportunity to use custom manufacturing, not yet in place, but clearly on the way with automated manufacturing tools. Still, this aspect is more controversial that the first two. The industrial revolution provided economy of scale by standardizing both product and production. To illustrate the possibility, examine the news story from May 7, 2012, The Printable House [4]. Danish architects Frederik Agdrup and Nicholas Bjorndal of Eentileen used CNC, (Computer Numerically Controlled), machines to construct a house. This allowed a lower cost approach to custom houses. This is only the potential start for increased custom products.

2   Background

Since the start of eCommerce and eBusiness, there have been efforts to formalize the design and operation of business. ECommerce has been concerned with the customer and online ordering and subsequent delivery of goods. This has matured to the state that almost all businesses need an online presence. The term eBusiness can include what was just describing as eCommerce, but it also includes B2B and that sense is what will be used here. B2B provides electronic interchange of data between businesses. This can include all kinds of business data, orders, bills, money transfers.

Ontology efforts have been applied to EBusiness [6, 7, 8]. These efforts have enabled an automated model of the business to be used to both understand and automate the operation of business, but also to provide a grammar for business to business operations.

In addition, some new business models have arisen. The automated auction model, pioneered by eBay, is common. Some of these are general selling market place, while others deal with specialized items. It did not take long for automated tools to be instituted to help the bidder keep up with the auctions. The typical auction bidding occurs in the last minute before the deadline. On could speculate the most of the auction period is there to generate interest and the real auction is the fast and furious last minutes.

Along with this, the rise of social media, and other data mining techniques, has produced the increased ability to record information and characterize the customers. This drives marketing to craft individualized marketing materials tailored to specific populations. This looks like another interesting dimension to the transformation of the business, but that is left for another paper or writer. Some of work [9] has been done to look at effective upsell techniques to automatically seek other offers to put to a buyer.

3   Formal Negotiation Process Description

As will be shown, negotiation is a fundamental part of the business model we propose.

In order to formalize the negotiation process, we draw on prior work. There are a number of available negotiation formalisms, the ones used here were chosen as a usable set. The base idea presented here is relatively independent of the specifics of the negotiation formalism.

3.1   A Negotiation Protocol

In Li, Su and Lam [2], a protocol for the negotiation process is given. They propose eight primitives as given in the table 1. This is a simple but complete protocol for conducting a negotiation.

Table 1. Negotiation protocol

CFP	Request a proposal
Propose a proposal	Send a proposal or a counter proposal that is acceptable to the sender
Reject proposal	Reject a proposal, with or without comment
Withdraw proposal	Withdraw a proposal
Accept proposal	Accept
Modify proposal	Modify either the CFP or the proposal
Acknowledge message	Acknowledgment for any message
Terminate negotiation	Unilaterally terminate the negotiation

The reference [1] also gives a state table for the operation of the protocol. Lau et. al [1] proceeds to develop a genetic algorithm as adaptive negotiation agents. The approach is of interest, but the actual negotiation calculation not critical to the approach here.

3.2   A Model for Proposals

In Lau et. al [1], a model for the proposals is given, and we present an overview of his model in this section. A proposal is an offer represented by a tuple. The offer is expressed with respect to a finite set of attributes $A = < a_1, a_2, \dots a_n >$. The offer is $o = < d_{a1}, d_{a2}, \dots d_{an} >$ where each d is a value or range from the domain of the attribute D_{ai}. The attributes can be viewed as the axis of a n dimensional space. The offers acceptable to a given agent are contained in a sub-region, or regions, of the space. Therefore the acceptable offers are a Cartesian product of the ranges of the attributes. Typical attributes include cost, delivery schedule, number of units, etc.

In order to evaluate a proposal, Lau et. al [1] proposed utility functions. U^a_p is a function on attribute a for agent p that maps on the interval [0,1]. This is used to normalize the evaluation. The sum of U^a_p over all a_i is one. The second function is U^D_p that gives a value over [0,1] for values in the domain of an attribute. The utility of an offer o is $U^o_p = \Sigma_a\ U^a_p \times U^D_p$, the sum over all attribute values multiplied by the normalization function.

Lau's [1] approach gives a method to compare offers and evaluate concessions on the offers. The function U^o_p produces a order over a set of offers. The parties to a negotiation will each have their own functions. As the negotiation proceeds the parties can keep a set of past offers and the function can be used to evaluate the possible counter offers by trying to maximize the value of the offer to the agent.

Lau et. al [1] proceeds to develop a genetic algorithm as adaptive negotiation agents. The approach is of interest, but the actual negotiation calculation not critical to the approach here.

3.3   The Full Problem

Most of the current work takes the view of a central negotiation manager (or agent). However, if the future of e-Commerce is fully realized, this is an over simplification of the operation of e-Commerce negotiation. The complexity will derive from two dimensions. First of all is a desire for speed. As the pace accelerates, the cost of not keeping up with the speed will clearly fall behind. The second factor is the interlocking of multiple negotiations that are interdependent.

The common view of systems architecture for enterprise architecture is expressed by the Open Group [3]. In this view, a central organizing principle is the idea of a single bottom line. This draws a boundary between the inside of the enterprise and the entities it interacts with. While this is a dominant view in current business practice, there is a possible tendency to break this down in the future. The mode supported by Amazon to act as a front end for many businesses presages as move to cooperate with competitors to increase sales for all. This could evolve into a more connected structure of business that cooperates in tightly networked businesses. For this reason, the work here tends to blur the boundaries of systems.

One interesting aspect of this model is the extent to which custom work can be part of the process. As an example of this, consider a clothing supplier that delivers custom fit clothing. In such a business the customer selects clothing style and provides sizes. The custom tailor shop has done this for centuries. In the new age, automated manufacture linked with an e business would extend this kind of custom clothing at nearer to mass production prices. In addition, options like materials and special features (i.e. custom pockets) are available with extra cost factors. This model of custom order can be extended into all kinds of products, were automated manufacture can reduce the costs.

The interlocking nature of the negotiations is an extension of the just in time concept in manufacturing. The goal of keeping inventory low, for tax and cash flow, required the careful management of the production rate, the inflow of materials and the shipping of product. This was started with a single plant and often one that produced a limited range of products. We believe that this model can be extended to a more complex and profitable system. In this model there are a set of interrelated negotiations. To examine the model we will first describe the basic components then progress to the interconnections.

The first component is the customer order system. This the logical first cause, since without the customer, in one form or another, there is no need to produce. The order system is concerned with delivery of the order and the price to be paid. It also must be prepared to negotiate over features. Some of the features may be custom, requiring design work. Another complication can be when multiple products need to be delivered together, rather than shipped as available.

Design is the element of the enterprise that must deal with new features or aspects that customize product.

Production is the organization that produces the product. In this model, multiple production facilities are assumed to exist. These facilities will compete for the production work. Since these are likely to be distributed, the choice is not purely based on capacity, but also delivery times from the facility to the customer. Also, a given customer order may be split between multiple production facilities, either because of capacity or because of specialization of the faculty for particular products.

Suppliers represent a number of external enterprises that provide raw materials for the production facility. Shipping represents the delivery of goods to an external entity that will move them to the customer. Shipping is a complex activity since cost and time are constraints on the shipping process. The successful enterprise needs to constantly evaluate the possible shipping options.

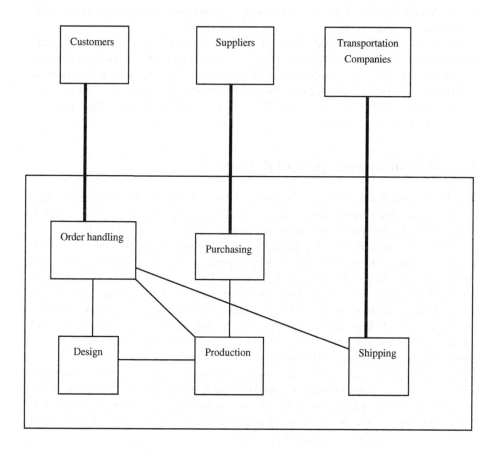

Fig. 1. De-centralized model of an enterprise

The view of the enterprise in Figure 1 is a de-centralized model. A model could be developed with a similar structure where a centralized coordinating manager exists. For this work, we will follow this more decentralized approach as it can be viewed as more flexible and dynamic. If a rapid business model is desired, then this de-centralized approach appears to have a great appeal. This view supports several ideas. First of all, allowing multiple design, production and order handling organizations that operate in concert. A different possible scenario is collaboration of independent, but trusted, businesses that work together to fulfill a customer need. The kind of flex-ible environment offered by e-business can enable this style of enterprise.

We further note that critical entities required for a functioning business are left out, notably there is no accounting and/or pricing activities. These were left out to keep the model smaller and, at the same time, reduce elements not related to negotiation activities. Pricing may be a negotiation activity, but for now we assume there are rather fixed business rules are known and applied across the enterprise. While these are omitted for the work here, in a functioning enterprise they do need to interact with the activities covered here.

Having the model described, the issues around the future evolution of the enter-prise can now be addressed. While the future described here may or may not come to be, we contend this is a possible state. This model is the extension of just in time manufacturing That is, an enterprise running at the speed of network enabled com-merce. To examine the idea and to establish why we conclude this is a viable future, a historic progression is examined.

4   The Future and Speed of Business

In order to argue for the future, we examine the past to see the trajectory and project into the future. The earliest businesses were probably conducted in the village market place. People bartered with each other with goods they had for other goods. Business in this example is business as a sideline to other activities in like framing or crafting. The next evolution would be business as an ongoing effort and the main focus on the business man's efforts. Business in this sense is the establishment of business as a separate occupation. In this early business model, people bartered for the goods avail-able and maybe for those to be delivered. As cities developed the market places grew, more was available from more businesses. The businesses become more complex and middlemen handled commerce in goods between suppliers and sellers. Shipping be-comes a business of its own as the cities grow.

As this world moved into the time of the industrial revolution, there was also a shift from custom items to standard goods. Prior to the industrial revolution, many goods were custom ordered, clothes made for the person, furniture with custom fea-tures, and many other items were treated as custom.   The industrial revolution lowered the cost of goods, but forced uniformity.

Another interesting shift was the catalog. In 1888, Sears produced a catalog for mail order. In America, the widely dispersed population of the west meant people were far from stores. The catalog allowed the widely spread population access to the goods available in the cities of the time. The world of Amazon is the electronic extension the old sears catalog. In the Amazon world, orders are processed by computer and order assembly is highly automated.

Amazon has added an interesting twist as well. Amazon sells not only from their own stock, but enables the buyer to purchase from any number of other sellers. Amazon has constructed a network of sellers that all use the same vehicle to reach a shared customer base. Amazon goods are orderable at any time and delivered within days.

Another enterprise shift of note is the change in entertainment businesses. We have moved from videotape in stores, to DVD's in stores to Netflix and on to streaming on demand video. 4G wireless networks are challenging the television cable systems. The displays are moving from TV's to iPad's, iPhones and Androids. All this allows rapid access to music, video and other possible forms of entertainment. A side effect of this is an increase in customization as well. No longer does a buyer have to buy an entire CD of music, just the songs desired. Likewise, television channels may be an idea of the past as buyers select the individual programs to view.

Shipping has also evolved in this time. It is now standard to manufacture anywhere and ship where needed. The control and precision which shippers now provide is surprising. Package tracking is a standard feature. Shipping times are much shorter. Many people buy a substantial portion of their goods online. This even extends to food in some areas.

We could also comment on the growth of social media as well as the increased ability to measure consumer behavior and preference.

In many ways, manufacturing has been a limiting factor to speed. The industrial revolution did speed up, as well as reduce the cost of, the production of goods. This was achieved by standardizing the goods produced. It also required design and set up work in order to have a production line with the tooling that reduces the skill of the workers and increases reliability and repeatability.

The current phase in manufacturing has been distribution. The earliest factories were self-contained. They even their own generated their own power. Over time, production has become distributed. Many complex parts are fabricated all over the world and shiped to final assembly points. Integrated circuits are often manufactured in several facilities (silicon foundries, packaging lines, testing etc.) and shipped between them.

A possible future growth in manufacturing is the increased use of automation to allow greater customization of the products. This is the actual inverse of the industrial revolution's standardization. While this is still a bit into the future, it does not seem to be that hard to achieve. There are many interesting manufacturing technologies that can be used as a basis for this step. Machining centers, robotic materials handlers, and the use of CAD design systems that can produce manufacturing data directly from a design tool are all current technologies. This can be extended into a fully automated manufacturing system. This has been demonstrated in some systems [4].

It does require a capable planning and control system. And it means more robotic factories to help produce the customized parts. An interesting idea, reported sometime back, is the custom book printer. In this system, the buyer selects a book and the book is custom printed and bound by an automated machine at the bookstore. Of course, this book production system may die, as electronic book readers become the preferred mode.

All these changes have a common effect; speed of transactions has been increasing. However, business processes still use old standard longer durations for bidding on proposals. This means that many business opportunities still have slow reaction times and require long planning intervals. The idea of a thirty-day bid period may become totally unacceptable.

What does this mean to businesses? Speed and flexibility will, increasingly, become a key to success. The ability to quickly respond to customer demand will become a competitive necessity. The flexibility needs to include the possibility to add new elements to the architecture of the business in an easy and agile approach. For example, if a customer wants custom features or items that are not in the current portfolio for the business, the ability to locate new designers, suppliers in a reliable and sound business practice can become key to landing the customer's order.

5   Enabling the High Speed Enterprise

To follow the ideas of the rapid speed requirements, the enterprise needs to become flexible and agile in decision making and planning. As this acceleration is increased, the speed will exceed human capability, especially with respect to enabling a large number of supplies, manufacturing facilities, and complex options to be considered. It is this issue that is the topic of the rest of this paper.

Given the distributed nature of the business, centralized planning does not offer the speed that will be needed in a global market with customized produces. In this model, not only will suppliers be able to bid on providing the sub assemblies and resources for a business to meet sales opportunities. The bidding model of [1] has the basic mechanism we need, however it needs some critical extensions. One extension is to add knowledge of the trust characteristics of the bidder based on their identity. There will be three dimensions to the negotiation based on identity; one will be inside the organization. One will be a highly trusted supplier and the last is suppliers in general.

The idea of negotiation within the organization may appear as a surprise at first. But it is a logical extension of speed. Human central planning is going to be too slow to operate in a high-speed enterprise. The roots of this switch are already evident. The on-line auction business already runs in an automated mode. The auction itself is attended by automated software. The bidders them selves are afforded automated tools to bid. It is almost impossible to win a highly contested auction with out the use of automated biding tools. The reality of the on-line auction is the bidding takes place in the end of the allotted time. While there are probably good psychological reasons for setting the time interval for the auction, with a more automated world those seem likely to be less necessary.

6   The Automated Bidder

In order to make the high-speed enterprise function the centralized planning can be replaced with an automated planner. The planner will be rules based, enabling the management to establish a set of business rules to order the work of the enterprise. The management of the enterprise becomes a process of examining the result via reports and data mining. These are used to tune the business rules. The planner system maybe a centralized single system, or may be a hierarchy of planners. The planners will need to have access to the current state of the business process and backlog. The planner system represents critical information needed by the bidding agents.

The central system will consist of the inventory system, the planner, the negotiation agent a bidding agent. These three will work in concert. The order can be filled by a combination of sources. The existing inventory, the planner can schedule production and parts or all of the order may be obtained from suppliers. In addition, parts need for the manufacturing or assembly process must be obtained, if they are not on hand.

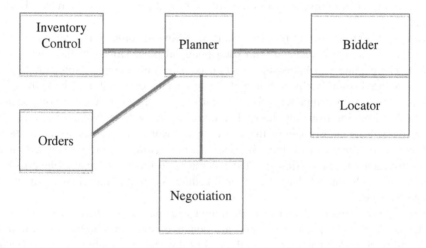

Fig. 2. The central system

The above figure shows the structure of these elements. Orders enter through the negotiation agent. The planner will be responsible for developing the answer to the request. The planner will check on inventory for the availability of both product and components for producing the product. The order system is consulted for components under order. The planner has access to the current committed production schedule and also needs the state of negations in progress. The planner needs to evaluate both the capacity to produce the product and the supplies r components required. The bidder is used to negotiation with other suppliers when components must be ordered. The suppliers can be internal or external. The external suppliers will be partitioned into preferred and regular. Preferred suppliers are those that enterprise has used before and may be one that have existing long-term arrangements in place.

The planner needs a rule base to implement the business rules for the business. These rules need to cover a range of issues. These include issues like pricing issues, determining fit of an opportunity to the business plans, choosing to sub-contract. These are clearly necessary issues to proper business operation. They are not the ones most interesting. There are several critical aspects to the operation of this complex from a speed of business perspective.

The planner must make a set of decisions when a potential order enters the system. The first being a basic go/no go business decision. There are several import aspects to this kind of decision. There will be an initial set of business issues alluded to above that evaluate the issues around the basic business fit. That is, is this a business opportunity that the business should attempt to engage. If this test is passed then the real work begins.

The first step is to determine a set of options. What tasks need to be used to fill the order. This may be a simple single option or may be set of options. The tasks come from varied set. It can use of the production facilities under the direct control of the entity. It can ask for production facilities from other division of the entity. It can sub-contract work from suppliers. It will include components from inventory, components transferred from inventory of other divisions and components s purchased from suppliers.

The planner must develop tentative plans for filling the order using the tasks. First of all there is a cost associated with each potential plan. However, the cost is not known. The planner will know the direct costs under direct control, for example the cost to use production facilities. Some of the component costs will also be knows, for example the cost on components from inventory, also the cost of some components may be fixed pricing from suppliers. He unknown costs will be evaluated to a range. The range will be an estimate of the best case and worst case values. Some of these will require negotiation with the other entities, but before starting he negotiation is seems logical to make an estimate of the cost range. At this point some of the possible plans may be eliminated due to cost or schedule issues that are simply out of an acceptable range.

The planner examines the current committed scheduled work. This is basic classical planner work. The plans developed are tentative, since this includes committed work and work under negotiation. At the end of this step, a possible set of plans and cost estimates are available. In addition they need to have a risk factor assigned. Risk is based on possible outcome of negotiations that must take place for supplies or contracts.

Based on the planner's output, there are several courses of action. First of all enough information may be available to make an offer or a tentative offer on the order. This will start a negotiation with the customer. Also, negations will be initiated with suppliers and contractors. This process will result is a series of negations that will increase the certainty of the costs. As offers and counter offers flow through the system, the planner can refine the possible plans and the final cost. This can be a highly dynamic process.

7   Opportunity Cost in Bidding

The one aspect not yet addressed is opportunity cost. The planner cannot work on single orders. The pace of business envisioned will require the planner to constantly rework the open negotiations in parallel. The orders can, and likely will, interfere with each other. At any time, there can be a mutually exclusive set of orders. This leads to the constant assessment of when to drop some orders. The drop decisions will be based on an assessment of opportunity. This can be the simplest case of the sure thing. That is an offer is accepted completely and choices are made to terminate some of the offers in play. The more complex case will be to choose to drop some negotiations early. This is based on assessment of opportunity. That is, the potential value of some offers change based on the sub negotiations, and the accepted work. This evaluation will depend on the business rules of the entity and the potential gain from each possible order.

The planner will have to interact with the bidding agent to obtain the products and services required for each order. This introduces several aspects that have to be put into a Lau [1] type negotiation system. First of all, contingent offers, the bids need to have expiration date. Thus a supplier is asked to bid and they include a time period that the bid is valid. If the set of linked negotiations can be completed in the time, then the offer will reach final acceptance. The process is to ask for bids, select the "best" set and then send an offer to the answer a customer request. If the customer accepts then the acceptances backward cascade through the chain of bids. At the same time, the planner may be working on customer orders that may over tax the ability of the business, so the planner will have to make choices between possible orders.

In addition, when an order is accepted, the resources required for teat order may interfere with other orders that are under bid. Therefore, acceptance not only generates the acceptance cascade through the bids that support the order, the planner must also determine if there are pending orders that need to be dropped and a cascade of rejections is executed.

To make the most profit, the value of the orders needs to be considered. The time element must come into the selection. The planner needs rules to determine how long to wait on accepting an order while a more valuable order is possible, but which will make the lesser value order impossible. The extension to the bidding and negotiation systems is to add time intervals. Orders are ranked by potential value and by time until the order must be answered. Potential value is estimated at first and then refined as bids are returned for the required products and services. The rank order is dynamic and will change as the bid progress.

When a lower valued order is complete, it can be accepted or delayed for higher value orders. The trade off must be made on the basis of time. There is also the risk that waiting too long will allow someone else to accept the order. This will be a critical business decision and one that will require some careful rule tuning. This decision logic will become a critical determiner for profit.

8   Conclusions

This chapter is highly speculative. It is a possible view of business in the future. The exciting aspect of this future is the reversal of the standardization of the indusial revolution. The future can become a world of custom items tailored for the individual. This does not mean that everything is custom, after all why customize things best left as commodities. The other aspect of high speed delivery and automated operation offers a lower cost point and business efficiency that seem hard to reject.

References

1. Lau, R.Y.K.: Adaptive Negotiation Agents for E-business. In: ICEC 2005 (2005)
2. Li, H., Su, S.Y.W., Lam, H.: On Automated e-Business Negotiations: Goal, Policy, Strategy and Plans of Decision and Action. Journal of Organizational Computing and Electronic Commerce 13(1), 1–29 (2006)
3. The Open Group, The Open Group Architectural framework (TOGAF) version 9. The Open Group, Reading, UK (2009), http://www.opengroup.org/togaf/
4. The Kurzweil Accelerating Intelligence newsletter,
   http://www.kurzweilai.net/in-denmark-a-printable-house?utm_source=KurzweilAI+Daily+Newsletter&utm_campaign=8179b0de4d-UA-946742-1&utm_medium=email (retrieved on May 7, 2012)
5. Turkle, S.: The Second Self: Computers and the Human Spirit. The MIT Press (2005) ISBN-13: 978-0262701112n Spirit
6. Ashraf, J., Cyganiak, R., O'Riain, S., Hadzic, M.: Open eBusiness Ontology Usage: Investigating Community Implementation of Good Relations. In: LDOW 2011, Hyderabad, India, March 29 (2011)
7. Osterwalder, A., Pigneur, Y.: An e-Business Model Ontology (eBMO) for Modeling eBusiness. In: The Bled EC Conference of the 15 th Bled eCommerce Conference (2002)
8. Singh, R., Iyer, L., Salam, A.F.: Semantic eBusiness. Int'l Journal on Semantic Web & Information Systems 1(1), 19–35 (2005)
9. Håkansson, A.: A Multi-Agent System with Negotiation Agents for e-Trading Products and Services. In: König, A., Dengel, A., Hinkelmann, K., Kise, K., Howlett, R.J., Jain, L.C. (eds.) KES 2011, Part IV. LNCS, vol. 6884, pp. 415–424. Springer, Heidelberg (2011)

Mathematical Models of Automated Auctions

Erol Gelenbe and Kumaara Velan

Intelligent Systems and Networks Group
Electrical and Electronic Engineering Department
Imperial College
London SW7 2BT, United Kingdom
{e.gelenbe,kumaara.velan}@imperial.ac.uk

Abstract. Decreasing computational costs and increasing connectivity via the Internet allows of using software entities to represent human counterparts in the digital marketplace, acting autonomously and yet guided by their design objectives to fulfil the interests of their owners to the best possible extent. In this chapter, we develop models of automated E-Commerce activities under the formal rule of automated auctions. Starting with small building blocks of interaction rules, we show how price formation may be predicted in these decision mechanisms using stochastic modelling theory. In the models, the auction proceeds in an ascending price order, and is represented as a random jump process that takes valuation from a state-space which represents the identity of the leading bidder as well as the attained price. Performance measures of a designated "Special Bidder" (SB) are obtained with respect to the rest of the auction participants. The bidder's optimal speed at which bids are placed (bid rate) in the auction with respect to the bid rate of the other bidders and the urgency of the seller to make a sale is derived.

1 Introduction

In this chapter, we develop models of automated E-Commerce activities under the formal rule of automated auctions. Decreasing computational costs and increasing connectivity via the Internet allows of using software entities to represent human counterparts in the digital marketplace, acting autonomously and yet guided by their design objectives to fulfil the interests of their owners to the best possible extent. Besides the obvious time-saving aspect on the part of the humans for automating the procedure, software agents are often more capable of performing real-time monitoring, computation and carrying out rule-based decisions accurately, and hence are able to achieve better efficiency than their human counterparts.

Among the economic institutions, auctions have well-defined protocols to ensure timely conclusions, and incorporate some notion of fairness by making any form of discriminate favouring among participants impossible on all aspects that do not contribute to the price. Starting with small building blocks of interaction rules, we show how the price formation may be predicted in these decision mechanisms, using stochastic modelling theory which is commonly used for analysing performance of queueing systems. We show how models of increasing complexity can be developed using simple first principles, and establish the probabilistic outcomes of the economic interactions. In the models, the auction proceeds in an ascending price order, and is represented as a

A. Håkansson & R. Hartung (Eds.): *Agent & Multi-Agent Syst. in Distrib. Syst.*, SCI 462, pp. 137–161.
DOI: 10.1007/978-3-642-35208-9_8   © Springer-Verlag Berlin Heidelberg 2013

random jump process that takes valuation from a countable state-space which denotes the attained price as well as the identity of the leading bidder.

The chapter begins with the problem description and motivation in the following section, in which the auction model is formalised generally with an arbitrary number of groups of bidders and an arbitrary number of bidders in each group. By introducing the assumption of exponential random variables for the time taken to submit bids by the bidders and the time taken to accept bids by the seller, we model the process as a continuous time Markov process in Section 3. In Section 4, we look at the reduced case of a single bidder called the Special Bidder (SB) against the other bidders and the seller, and introduce the measures of interests that determine its performance in the auction. Following that, the same case is studied with an approximate model 5, which allows closed form expressions for the performance measures. Using these results, an optimisation problem faced by the SB is formulated in Section 6, in which it balances its two conflicting measures of interests. Section 7 extends the previous model to the case where the SB may have valuations that are distributed non identically to the valuation of the other bidders. Finally, we conclude the chapter with remarks on the implications of the results and discuss some interesting questions.

2   Ascending Auction with Discrete Increments

We consider an ascending auction in which successive bid offers need to be higher than the previous by some discrete valuation. Suppose N groups of bidders participate in an auction, and the i-th group consists of n_i bidders for $i = 1, \ldots, N$. Further, let each bidder in group i be indexed with $k_i = 1, \ldots, n_i$. The good being sold has a maximum valuation v, beyond which no bidders will make offers.

The auction then proceeds with the sequence of events $0 < T_1 < T_2 \cdots < T_Z$, where the auction starts at 0, closes at T_Z, and the rest of the times represent the events of incoming bids. Each bidder will naturally take some time to consider the status of the auction before deciding to make an offer. Let $B_{i,k_i,j}$ be the amount of time taken by the k_i-th bidder from group i for placing the j–th bid. The bid increment by the k_i-th bidder from group i for the j–th bid is $I_{i,k_i,j}$, and the actual j–th bid is L_j. The sequence of times representing events of incoming bids is generated with the rule

$$T_1 = \min_{i,k_i} \left[B_{i,k_i,j} \right] \tag{1}$$

$$T_{j+1} = T_j + \min_{i,k_i} \left[B_{i,k_i,j+1} \right] \tag{2}$$

The bidder who is leading after the j–th bid is the one who placed his bid in the shortest time $(w, k_w) = \arg\min_{i,k_i} [B_{i,k_i,j}]$, and his bid increment will determine the valuation of the current bid offer $L_j = L_{j-1} + I_{w,k_w,j}$. After the j–th bid, the seller spends some time considering the offer and this quantity is represented as D_j. Following a simple rule, the auction concludes after the j–th bid if $T_j + D_j < T_{j+1}$, i.e. the seller decides to accept an offer before the next offer is received. Until this rule is triggered, the auction continues soliciting bids from participants. If $Z = \min[j : T_j + D_j < T_{j+1}]$ then the winner is $(w, k_w) = \arg\min_{i,k_i}[B_{i,k_i,Z}]$ at price L_Z which is also the seller's income. The total duration of the auction is $T_D = T_Z + D_Z + R$. The parameter R is the period of time before the next cycle of auction is started, and we model it with a general distribution function $R(t) = P[R < t]$. In this sequence the whole process is repeated indefinitely and independently of the previous.

In each auction, the payoff earned by the bidders is

$$\pi_{i,k_i} = \begin{cases} v - L_Z, & \text{if } (i, k_i) = \arg\min_{i,k_i}[B_{i,k_i,Z}]; \\ 0, & \text{otherwise} . \end{cases} \tag{3}$$

and the income rate earned by the seller is $I_S = \frac{L_Z}{T_D}$.

If the above variables are random, the following form of transition probabilities allows us to characterise the auction as a Markov process:

$$P\{B_{i,k_i,j+1} \le t, I_{i,k_i,j+1} \le m, D_{j+1} \le d | L_j = b\} \tag{4}$$
$$= P\{B_{i,k_i,j+1} \le t | L_j = b\} \cdot P\{I_{i,k_i,j+1} \le m | L_j = b\} \cdot P\{D_{j+1} \le d | L_j = b\}, \tag{5}$$

and this allows for the bid times, bid increments, and decision delays to be independent of each other but may easily depend on the valuation of the most recent bid b. Thus, the bidder's behaviour is independent of the seller's, but both are determined by the price attained. In addition, the bidders also behave independently of each other, or

$$P\left(\bigcap_{i,k_i}\{B_{i,k_i,j} \le t_{i,k_i,j}\}\right) = \prod_{i,k_i} P(\{B_{i,k_i,j} \le t_{i,k_i,j}\}) \tag{6}$$

and

$$P\left(\bigcap_{i,k_i}\{I_{i,k_i,j} \le m_{i,k_i,j}\}\right) = \prod_{i,k_i} P(\{I_{i,k_i,j} \le m_{i,k_i,j}\}). \tag{7}$$

3   The Case with Unit Increments and Exponential Bidding and Decision Times

If we assume unit price increments $I_{i,k_i,j} = 1$, that is subsequent bid offers always exceed the previous by a unit, we arrive at a special case of the described model which was studied in [13]. In addition the variables $B_{i,k_i,j}$ are replaced with exponentially distributed random variables with parameter β_i, thus bidders from the same group have independent periods of time drawn from the same common distribution. Also, for the present, this parameter is not dependent on the bid position j, although models with bid dependent parameters can also be studied, as we have done in [12].

The seller's decision times are modelled with exponential random variables with parameter δ. With these assumptions, the model is reduced to a discrete state system, where the state contains information on the group identity of the leading bidder and the leading bid for the item. The process is a continuous-time Markov process $\{X_t : t \ge 0\}$ with state space

$$X_t \in Y = \{0, B(i,l), A(i,l) : 1 \le l \le v, 1 \le i \le k\}, \tag{8}$$

which has the size of $2vk+1$. The process will be in state $B(i,l)$ when the highest bid is held by any single bidder from the i–th group at the price l, and it will be in state $A(i,l)$ when the auction concludes with a sale to a bidder from the i–th group at the price l and waits for the next restart. The long-run stationary probabilities for any state $x \in Y$ is defined as

$$P(x) = \lim_{t \to \infty} P\{ X_t = x \}. \tag{9}$$

Then the stationary probabilities will satisfy the following system equations:

$$P(B(i,1)) \left[\sum_{j=1,j\neq i}^{k} n_j\beta_j + (n_i - 1)\beta_i + \delta \right] = P(0)n_i\beta_i, \quad i = 1,\ldots,k, \tag{10}$$

$$P(B(i,l)) \left[\sum_{j=1,j\neq i}^{k} n_j\beta_j + (n_i - 1)\beta_i + \delta \right]$$

$$= \sum_{j=1,j\neq i}^{k} P(B(j,l-1))n_i\beta_i + P(B(i,l-1))(n_i - 1)\beta_i,$$

$$2 \leq l \leq v-1, \quad i = 1,\ldots,k, \tag{11}$$

$$P(B(i,v))\delta = \sum_{j=1,j\neq i}^{k} P(B(j,v-1))n_i\beta_i + P(B(i,v-1))(n_i - 1)\beta_i,$$

$$i = 1,\ldots,k, \tag{12}$$

$$P(A(i,l))r = \delta P(B(i,l)), \quad 1 \leq l \leq v, \quad i = 1,\ldots,k, \tag{13}$$

$$P(0)\sum_{i=1}^{k} n_i\beta_i = r\sum_{i=1}^{k}\sum_{l=1}^{v} P(A(i,l)), \tag{14}$$

$$P(0) + \sum_{i=1}^{k}\sum_{l=1}^{v} (P(A(i,l)) + P(B(i,l))) = 1. \tag{15}$$

In the first equation, we observe that the total flow of leaving the state $B(i,1)$–which corresponds to a group i bidder holding the highest bid offer at price 1–is the summation of bidding intensities of all members from other groups, together with the bidding intensity of all but the current highest bidder of group i, and the rate at which the offer is accepted by the seller. If the long-run stationary probabilities exist, this flow will be balanced with the total flow at which the system enters this state, i.e. cumulative bidding rate of all in group i when the system has just started with price zero.

For each state $B(i,l)$ $1 \leq l \leq v-1$, while the ways in which the process leaves the state remains the same as in the former case, there are different possibilities of entering it: when the price is one unit less and the highest bidder is one from the other groups, all bidders in group i will bid; if the highest bid is owned by one of their own, however, only the remaining (not the highest bidder) players will bid. The same reasoning applies for the incoming flow at state $B(i,v)$, but it is a different case for the outgoing flows: as no bidders from any of the groups will be willing to raise the price beyond the maximum, the only exiting transitions from these states is contributed by the selling rates.

The remaining equations capture the following features of the system as described. In (13) the total flow into the sale state at price l to a participant of group i, $A(i,l)$, originates from sale events from the corresponding highest-offer states $B(i,l)$, and must be balanced by the outgoing flow due to the auction restarting for the subsequent cycle. Since the auction restarting rate is invariant, and is independent of the concluding sale states, the combined flow from all these states into state 0 i.e. when the auction is waiting for bids, is balanced by the rate at which all the bidders collectively make the first bid (see 14). Finally, (15) applies the law of total probability for the system.

It should be noted that the grouping of bidders does not in any way imply collective selfish behaviour among group members; a group member will always raise the offer price if he is not winning, even if one of his own group members holds the highest offer. The grouping is simply a convenient mathematical abstraction to view collectively a set of bidders who share identical statistical properties, such as bidding rates in this case.

Some general results of this system can be derived easily. The total stationary probabilities of all the sale states, and the highest-offer states are

$$\sum_{i=1}^{k}\sum_{l=1}^{v} P(A(i,l)) = P(0)\frac{\sum_{i=1}^{k} n_i \beta_i}{r}, \quad \text{and} \tag{16}$$

$$\sum_{i=1}^{k}\sum_{l=1}^{v} P(B(i,l)) = P(0)\frac{\sum_{i=1}^{k} n_i \beta_i}{\delta}, \tag{17}$$

respectively. Consequently, after some algebra, we obtain

$$P(0) = \frac{r\delta}{r\delta + (r+\delta)(\sum_{i=1}^{k} n_i \beta_i)}. \tag{18}$$

In general, there are no straightforward expressions for the stationary probabilities $P(A(i,l))$ and $P(B(i,l))$. Nevertheless, numerical values can be computed efficiently for any arbitrary setting of parameters. We can make two observations regarding how the probabilities relate to each other: for any i,l, $P(B(i,l))$ is necessary and sufficient to compute $P(A(i,l))$; evaluating $P(B(i,l))$ does not require $P(B(j,m))$ for all $m \geq l$ and $m < l - 1$. Hence the solutions can be computed in the following order:

1. Find $P(0)$.
2. In a sequential order for each $l = 1,\ldots,v$, solve $P(B(i,l))$ for all i.
3. Solve for all $P(A(i,l))$.

3.1 Performance Measures

The performance measures pertaining to each bidder belonging to any group can be defined and evaluated in a straightforward fashion. The average duration of an auction cycle is

$$\tau = \frac{P(0)^{-1}}{\sum_{i=1}^{k} n_i \beta_i} = \frac{r\delta + (r+\delta)(\sum_{i=1}^{k} n_i \beta_i)}{r\delta \sum_{i=1}^{k} n_i \beta_i}. \tag{19}$$

The probability that a particular single bidder in group i wins is

$$\pi(i) = \frac{\sum_{l=1}^{l=v} P(A(i,l))}{n_i \sum_{i=1}^{i=k} \sum_{l=1}^{l=v} P(A(i,l))}. \tag{20}$$

The expected savings a bidder in group i generates with respect to the maximum that it is willing to pay *given that it wins* can be expressed as

$$\phi(i) = \frac{\sum_{l=1}^{l=v}(v-l)P(A(i,l))}{n_i \sum_{l=1}^{l=v} P(A(i,l))}. \tag{21}$$

4   A Single Bidder Against the Market

One useful way of exploiting the model is to ask how well can a single bidder who arrives at the auction and observes the ongoing activities of bidding and selling do against the rest of the participants. We designate this single bidder as the special bidder "SB" who is against the other bidders making offers to the seller. Effectively this corresponds to the case of two groups: the first group consists of the SB alone bidding at rate β_1, and the second group consists of n_2 bidders each bidding at rate β_2. With this the rest of participants are aggregated in a common group and allowed to have a common bidding rate. Although this restricts the previous general model in that the other bidders are not allowed to have independent parameters, we will see that this restriction helps us find a closed form solution for the stationary probabilities, which is not possible without the restriction.

In the present case, the state-space (8) is reduced to

$$X_t \in Y = \{0, B(1,l), B(2,l), A(1,l), A(2,l) : 1 \le l \le v\}, \tag{22}$$

in which $B(1,l)$ and $A(1,l)$ correspond to, respectively, states when the SB is leading with the highest bid in auction and when the SB has actually won the item at price l. Similarly $B(2,l)$ represents the states when one of the other bidders is leading and $A(2,l)$ the states when one of the other bidders has won the item at price l. The system equations satisfied by the stationary probabilities, (10) till (15), simplify to the following:

$$P(B(2,1))((n_2-1)\beta_2+\beta_1+\delta) = n_2\beta_2 P(0), \tag{23}$$
$$P(B(2,l))((n_2-1)\beta_2+\beta_1+\delta) = (n_2-1)\beta_2 P(B(2,l-1))$$
$$+ n_2\beta_2 P(B(1,l-1)), \quad 2 \le l \le v-1,$$
$$P(B(2,v))\delta = (n_2-1)\beta_2 P(B(2,v-1)) + n_2\beta_2 P(B(1,v-1)),$$
$$P(A(2,l))r = \delta P(B(2,l)), \quad 1 \le l \le v,$$
$$P(B(1,1))(n_2\beta_2+\delta) = \beta_1 P(0),$$

$$P(B(1,l))(n_2\beta_2 + \delta) = \beta_1 P(B(2,l-1)), \quad 2 \le l \le v-1,$$
$$P(B(1,v))\delta = \beta_1 P(B(2,v-1)),$$
$$P(A(1,l))r = \delta P(B(1,l)), \quad 1 \le l \le v,$$
$$P(0)(\beta_1 + n_2\beta_2) = r \sum_{U=1,2} \sum_{l=1}^{v} P(A(U,l)),$$
$$1 = P(0) + \sum_{U=1,2} \sum_{l=1}^{v} \left[P(B(U,l)) + P(A(U,l)) \right].$$

Unlike the previous, this case has an analytical solution. First introduce the sequences $H(l)$ and $G(l)$,

$$H(l) = \begin{cases} \dfrac{n_2\beta_2}{\beta_1 + (n_2-1)\beta_2 + \delta}, & l=1 \\[2ex] \dfrac{(n_2-1)\beta_2}{\beta_1 + (n_2-1)\beta_2 + \delta}H(l-1) \\[2ex] \quad + \dfrac{n_2\beta_2}{\beta_1 + (n_2-1)\beta_2 + \delta}G(l-1), & 2 \le l \le v-1 \\[2ex] \dfrac{(n_2-1)\beta_2}{\delta}H(l-1) + \dfrac{n_2\beta_2}{\delta}G(l-1), & l=v \end{cases} \qquad (24)$$

$$G(l) = \begin{cases} \dfrac{\beta_1}{n_2\beta_2 + \delta}, & l=1 \\[2ex] \dfrac{\beta_1}{n_2\beta_2 + \delta}H(l-1), & 2 \le l \le v-1 \\[2ex] \dfrac{\beta_1}{\delta}H(l-1), & l=v \end{cases}$$

so that the stationary probabilities can be written as:

$$P(B(2,l)) = H(l)P(0), \qquad (25)$$
$$P(B(1,l)) = G(l)P(0),$$
$$P(A(2,l)) = \frac{\delta}{r}H(l)P(0),$$
$$P(A(1,l)) = \frac{\delta}{r}G(l)P(0).$$

The stationary probability for state 0, which represents the auction having restarted and waiting to receive the first bid, is a special case of (18) involving only two groups:

$$P(0) = \frac{r\delta}{r\delta + (r+\delta)(\beta_1 + n_2\beta_2)}. \qquad (26)$$

The analytical solution for $H(l)$ where $1 \le l \le v-1$ is obtained in the following. First, it is helpful to define the coefficients in (24)

$$\alpha_1 = \frac{n_2\beta_2}{\beta_1 + (n_2-1)\beta_2 + \delta}, \tag{27}$$

$$\alpha_2 = \frac{(n_2-1)\beta_2}{\beta_1 + (n_2-1)\beta_2 + \delta},$$

$$\alpha_3 = \frac{(n_2-1)\beta_2}{\delta},$$

$$\alpha_4 = \frac{n_2\beta_2}{\delta},$$

$$\alpha_5 = \frac{\beta_1}{n_2\beta_2 + \delta},$$

$$\alpha_6 = \frac{\beta_1}{\delta}.$$

Observing the recurrence relations in $H(l)$, and by substituting $G(l-1)$ with its representation as a function of $H(l-2)$, we can write for all $3 \le l \le v-1$

$$H(l) = \alpha_2 H(l-1) + \alpha_1\alpha_5 H(l-2), \tag{28}$$

in which the initial values are $H(1) = \alpha_1$ and $H(2) = \alpha_1(\alpha_2 + \alpha_5)$. This recurrence equation is then solved for the roots

$$R_1 = \frac{1}{2}\left[\alpha_2 + \sqrt{\alpha_2^2 + 4\alpha_1\alpha_5}\right] \quad \text{and}$$

$$R_2 = \frac{1}{2}\left[\alpha_2 - \sqrt{\alpha_2^2 + 4\alpha_1\alpha_5}\right].$$

The analytical solution for the sequence $H(l)$ can be expressed solely in terms of the roots and the coefficients.

$$H(l) = \frac{1}{2(R_1-R_2)}\Big[(-\alpha_2 + 2\alpha_1 + R_1 - R_2)R_1^l$$
$$+ (\alpha_2 - 2\alpha_1 + R_1 - R_2)R_2^l\Big], \quad 1 \le l \le v-1. \tag{29}$$

The solution at the boundary $l = v$ involves a different set of coefficients

$$H(v) = \alpha_3 H(v-1) + \alpha_4\alpha_5 H(v-2). \tag{30}$$

Now we can easily substitute $G(l)$ with expressions for $H(l-1)$ to obtain

$$G(1) = \alpha_5, \tag{31}$$

$$G(l) = \frac{\alpha_5}{2(R_1-R_2)}\Big[(-\alpha_2 + 2\alpha_1 + R_1 - R_2)R_1^{l-1}$$
$$+ (\alpha_2 - 2\alpha_1 + R_1 - R_2)R_2^{l-1}\Big], \quad 2 \le l \le v-1, \quad \text{and}$$

$$G(v) = \alpha_6 H(v-1).$$

With the approximation, the set of balance equations satisfied by the stationary probabilities (23) become:

$$P(B(2,1)) = \rho P(0), \tag{37}$$
$$P(B(2,l)) = \rho P(l-1), \quad 2 \leq l \leq v-1,$$
$$P(B(2,v))\delta = n\beta_2 P(v-1),$$
$$P(A(2,l))r = \delta P(B(2,l)), \quad 1 \leq l \leq v,$$

and the equations pertaining to the SB become

$$P(B(1,1)) = \mu P(0), \tag{38}$$
$$P(B(1,l)) = \mu P(B(2,l-1)), \quad 2 \leq l \leq v-1,$$
$$P(B(1,v))\delta = \beta P(B(2,v-1)),$$
$$P(A(1,l))r = \delta P(B(1,l)), \quad 1 \leq l \leq v,$$

while the last two equations in (23), which pertain to $P(0)$, remain unchanged.

We proceed to solve the set of equations in the same manner. Despite the changes to the model, the valuation for $P(0)$ remains the same as (26). For the rest of the stationary probabilities, first define the sequences

$$H(l) = \begin{cases} \rho, & l=1; \\ \rho\left[H(l-1)+G(l-1)\right], & 2 \leq l \leq v-1; \\ \dfrac{n\beta_2}{\delta}\left[H(l-1)+G(l-1)\right], & l=v, \end{cases}$$

and

$$G(l) = \begin{cases} \mu, & l=1; \\ \mu H(l-1), & 2 \leq l \leq v-1; \\ \dfrac{\beta_1}{\delta}H(l-1), & l=v. \end{cases}$$

The recurrence relation in $H(l)$ is obvious:

$$H(l) = \rho H(l-1) + \mu\rho H(l-2), \quad 3 \leq l \leq v-1.$$

Let R_1 and R_2 be the roots of this equation, then:

$$R_1 = \frac{1}{2}\left[\rho + \sqrt{\rho^2 + 4\rho\mu}\,\right] \quad \text{and} \quad R_2 = \frac{1}{2}\left[\rho - \sqrt{\rho^2 + 4\rho\mu}\,\right].$$

As a consequence, we can write the sequences $H(l)$ in closed form, thus allowing expressions for the stationary probabilities as well:

$$P(B(2,l)) = P(0)\frac{R_1^{l+1} - R_2^{l+1}}{R_1 - R_2}, \quad 1 \leq l \leq v-1,$$

$$P(A(2,l)) = P(0)\left[\frac{R_1^{l+1} - R_2^{l+1}}{R_1 - R_2}\right]\frac{\delta}{r}, \quad 1 \leq l \leq v-1.$$

At the boundary $l = v$, the solution involves a different set of coefficients

$$P(B(2,v)) = P(0) \left[\frac{R_1^v - R_2^v + \mu(R_1^{v-1} - R_2^{v-1})}{R_1 - R_2} \right] \frac{n\beta_2}{\delta}$$

$$P(A(2,v)) = P(0) \left[\frac{R_1^v - R_2^v + \mu(R_1^{v-1} - R_2^{v-1})}{R_1 - R_2} \right] \frac{n\beta_2}{r}.$$

The solutions relating to the SB are

$$P(B(1,1)) = \mu P(0), \tag{39}$$

$$P(B(1,l)) = \mu P(0) \left[\frac{R_1^l - R_2^l}{R_1 - R_2} \right], \quad 2 \le l \le v - 1,$$

$$P(B(1,v)) = P(0) \left[\frac{R_1^v - R_2^v + \mu(R_1^{v-1} - R_2^{v-1})}{R_1 - R_2} \right] \frac{n\beta_2}{\delta},$$

and

$$P(A(1,1)) = P(0)\frac{\mu\delta}{r}, \tag{40}$$

$$P(A(1,l)) = P(0) \left[\frac{R_1^l - R_2^l}{R_1 - R_2} \right] \frac{\mu\delta}{r}, \quad 2 \le l \le v - 1,$$

$$P(A(1,v)) = P(0) \left[\frac{R_1^v - R_2^v + \mu(R_1^{v-1} - R_2^{v-1})}{R_1 - R_2} \right] \frac{n\beta_2\mu}{r}.$$

Similar to the previous case, because $\rho > 0$, $\mu > 0$, and $\rho^2 + 4\rho\mu > 0$, it follows that $R_1 - R_2 = \sqrt{\rho^2 + 4\rho\mu} > 0$; thus the solutions exist and take real valuations.

5.1   Closed Form Expressions for the Performance Measures

Using (40), the sum of probabilities for all cases where the SB wins the item is easily obtained

$$\sum_{l=1}^{v} P(A(1,l)) = P(0) \left[\frac{\delta\mu}{r} + \frac{\delta\mu}{r(R_1 - R_2)} \left[\frac{R_1^v - R_1^2}{R_1 - 1} - \frac{R_2^v - R_2^2}{R_2 - 1} \right] \right.$$
$$\left. + \frac{\beta_1}{r} \left[\frac{R_1^v - R_2^v}{R_1 - R_2} \right] \right]. \tag{41}$$

Also, the summation weighted by the winning price

$$\sum_{l=1}^{v} lP(A(1,l)) = P(0) \left[\frac{\delta\mu}{r} + \frac{v\beta_1}{r} \left[\frac{R_1^v - R_2^v}{R_1 - R_2} \right] \right.$$
$$+ \frac{\delta\mu}{r} \frac{1}{R_1 - R_2} \left[\frac{(v-1)R_1^{v+1} - vR_1^v - R_1^3 + 2R_1^2}{(R_1 - 1)^2} \right.$$
$$\left. \left. - \frac{(v-1)R_2^{v+1} - vR_2^v - R_2^3 + 2R_2^2}{(R_2 - 1)^2} \right] \right]. \tag{42}$$

As a consequence, the probability that the SB wins the good rather than one of the others (33) can be expressed in closed form

$$\pi(1) = \left[\frac{r}{\beta_1 + n_2\beta_2}\right] \cdot \left[\frac{\delta\mu}{r} + \frac{\delta\mu}{r(R_1 - R_2)}\left[\frac{R_1^v - R_1^2}{R_1 - 1} - \frac{R_2^v - R_2^2}{R_2 - 1}\right] + \frac{\beta_1}{r}\left[\frac{R_1^v - R_2^v}{R_1 - R_2}\right]\right].$$

(43)

The time that is spent by the SB waiting before a successful purchase is

$$\psi = \frac{r\delta + (r + \delta)(\beta_1 + n_2\beta_2)}{r^2\delta\left[\frac{\delta\mu}{r} + \frac{\delta\mu}{r(R_1 - R_2)}\left[\frac{R_1^v - R_1^2}{R_1 - 1} - \frac{R_2^v - R_2^2}{R_2 - 1}\right] + \frac{\beta_1}{r}\left[\frac{R_1^v - R_2^v}{R_1 - R_2}\right]\right]}$$

(44)

Also, the average savings the SB makes on a purchase given that it wins is obtained in a somewhat lenghtier expression

$$\phi = \frac{v\sum_{l=1}^{v} P(A(1,l)) - \sum_{l=1}^{v} lP(A(1,l))}{\sum_{l=1}^{v} P(A(1,l))}$$

$$= \frac{\frac{\delta\mu(v-1)}{r} + \frac{\delta\mu}{r(R_1 - R_2)}Y}{\frac{\delta\mu}{r} + \frac{\delta\mu}{r(R_1 - R_2)}\left[\frac{R_1^v - R_1^2}{R_1 - 1} - \frac{R_2^v - R_2^2}{R_2 - 1}\right] + \frac{\beta_1}{r}\left[\frac{R_1^v - R_2^v}{R_1 - R_2}\right]},$$

(45)

where

$$Y = \frac{v^2 R_1^{2v} - 2v^2 R_1^{v+2} - (v-1)R_1^{v+1} + vR_1^v + v^2 R_1^4 + R_1^3 - 2R_1^2}{(R_1 - 1)^2}$$

$$- \frac{v^2 R_2^{2v} - 2v^2 R_2^{v+2} - (v-1)R_2^{v+1} + vR_2^v + v^2 R_2^4 + R_2^3 - 2R_2^2}{(R_2 - 1)^2}.$$

Regarding the seller's interests, an analytical expression for its income rate S (36) is possible but is lengthy and unwieldy. However, using results from [3], we can find bounds for this quantity. Note that in [3] all bidders are aggregated into a single pool that generates bid offers; thus bidders have identical bidding rates. In our present case, bidders have different bidding rates, but the highest bid rate that the seller can expect is $\beta_1 + n_2\beta_2$ which only occurs when the auction has just restarted, i.e. at price 0. At any state after this, surely one bidder from the pool will restrain from bidding, since he holds the leading bid, resulting in a lower cumulative bid rate. Therefore this case gives us the upper bound on the income rate for the seller.

$$S < \left[\frac{r(\beta_1 + n_2\beta_2)(\beta_1 + n_2\beta_2 + \delta)}{(\beta_1 + n_2\beta_2)(r + \delta) + r\delta}\right]\left[1 - \left[\frac{\beta_1 + n_2\beta_2}{\beta_1 + n_2\beta_2 + \delta}\right]^v\right]$$

(46)

Following the same argument, suppose the bid rate that the seller can expect is independent of the price attained i.e. whether $l = 0$ or $l > 0$, then the lowest bid rate that can be expected at any time is $x = \min\{\beta_1 + (n_2 - 1)\beta_2, n_2\beta_2\}$, giving us the lower bound

$$S > \left[\frac{rx(x + \delta)}{x(r + \delta) + r\delta}\right]\left[1 - \left[\frac{x}{x + \delta}\right]^v\right].$$

(47)

Figure 1 shows a numerical example for the seller's income rate S computed numerically from (36) with the upper and lower bounds from closed form expressions (46) and (47).

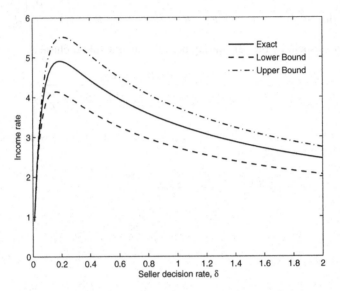

Fig. 1. The seller's income per unit time against its decision rate δ. The bounds are computed with closed form expressions while the exact valuations are computed numerically. Other parameters: $\beta_1 = 2$, $\beta_2 = 0.5$, $r = 1$, $n = 10$

5.2 Numerical Examples

The results (43) to (45) allow us to evaluate the measures of interest directly, in a single computation, rather than the individual term evaluations and sequential summations that equations (33) to (35) entail. Obviously the cost is the accuracy of the evaluations, but as we can see in the examples in Figure 2 for the expected time to win and in Figure 3 for the expected payoff, the approximations are close to the exact valuations. Both quantities decrease with the SB's bid rate β_1, but the time to win decreases much more rapidly with β_1. Also, as might be intuitively expected, the time to win increases and the payoff decreases with increasing competition from the other bidders in the form of higher bid rates β_2.

6 Bidder with Time Constraints

Our studies have shown that the two quantities of interest for the SB, the average time spent in waiting before a successful purchase ψ and the average payoff or savings made with respect to the SB's maximum valuation of the good ϕ, have conflicting response to increasing the speed at which bids are submitted β_1. When the SB is quick at submitting bids, i.e. high β_1, it can usually expect to win the item sooner which corresponds to a small valuations for ψ. However, a high bid rate also means that the SB, on average,

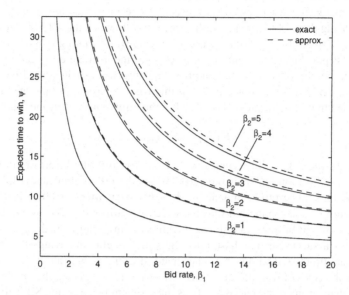

Fig. 2. The SB's expected time to win against bidding rate β_1, comparing exact solutions with approximation results for differing levels of the other's bidding rate. Other parameters are $\delta = 0.5$, $r = 1$, $n = 10$, and $V \sim U(80, 100)$.

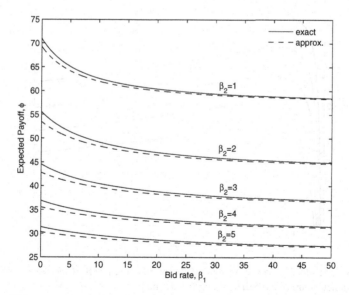

Fig. 3. The SB's expected payoff against bidding rate β_1, comparing exact solutions with approximation results for differing levels of the other's bidding rate. Other parameters are $\delta = 0.5$, $r = 1$, $n = 10$, and $V \sim U(80, 100)$.

does not strike the best deal on the price and thus the quantity ϕ reduces in value. A compromise between these opposing interests is the best way forward.

In formalising the problem, suppose the urgency with which the SB needs to procure the item can be quantified with the cost that it incurs while waiting in the auction. Let c_w be the cost per unit time spent. Then $c_w\psi$ gives the average cost paid by the SB for waiting to purchase the item; for simplicity assume the costs increase linearly with time. Consequently, the difference between the average payoff and the average cost, both expressed in unit money, can be defined as the utility enjoyed by the bidder:

$$U(\beta_1) = \phi - c_w\psi \tag{48}$$

Figure 6 traces the SB's utility against its bid rate for differing levels of competition from the others' in terms of their bid rates. When the others' bid rate is higher, the SB can expect a smaller utility in general. However, even after this effect is taken into account, there exists a unique bid rate for which the utility enjoyed by the bidder is maximised. If the SB bids at a rate slower than this optimal valuation, the costs it incurs in waiting for a successful purchase outweighs the benefits in savings that it makes in striking an attractive deal. On the other hand, if a higher bid rate is exercised, the reduction in the costs of waiting for the SB is neutralised and outweighed by the loss in the savings that it can enjoy otherwise. Thus the problem faced by the SB is to find the optimal bid rate that balances its costs against its payoffs.

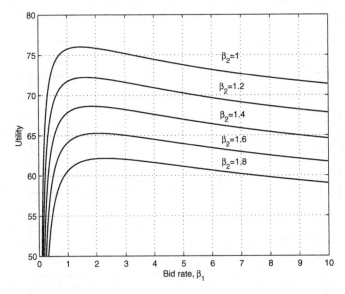

Fig. 4. The SB's utility, $U = \phi - c_w\psi$, against β_1 for differing levels of the other's bidding rate β_2. Other parameters are $\delta = 0.5$, $n = 10$, $c_w = 0.1$, and $r = 1$.

Replacing the quantities ψ and ϕ with the expressions (44) and (45), the utility for the SB is

$$U(\beta_1) = v - \frac{\frac{\delta\mu}{r} + \frac{v\beta_1}{r}\left[\frac{R_1^v - R_2^v}{R_1 - R_2}\right] + \frac{\delta\mu}{r(R_1 - R_2)}\left[\frac{(v-1)R_1^{v+1} - vR_1^v - R_1^3 + 2R_1^2}{(R_1 - 1)^2} - \frac{(v-1)R_2^{v+1} - vR_2^v - R_2^3 + 2R_2^2}{(R_2 - 1)^2}\right]}{\frac{\delta\mu}{r} + \frac{\delta\mu}{r(R_1 - R_2)}\left[\frac{R_1^v - R_1^2}{R_1 - 1} - \frac{R_2^v - R_2^2}{R_2 - 1}\right] + \frac{\beta_1}{r}\left[\frac{R_1^v - R_2^v}{R_1 - R_2}\right]}$$
$$- \frac{c_w(r\delta + (r+\delta)(\beta_1 + n_2\beta_2))}{r^2\delta\left[\frac{\delta\mu}{r} + \frac{\delta\mu}{r(R_1 - R_2)}\left[\frac{R_1^v - R_1^2}{R_1 - 1} - \frac{R_2^v - R_2^2}{R_2 - 1}\right] + \frac{\beta_1}{r}\left[\frac{R_1^v - R_2^v}{R_1 - R_2}\right]\right]}. \tag{49}$$

We must remind ourselves that this expression yields an approximation, and not the exact valuation of the true utility which can be computed numerically using the results obtained in Section 4.

Owing to the ability to express the utility in closed form (49), and by applying a further approximation taking a large valuation $v \to \infty$, we have an expression that approximates the solution to the optimisation problem:

$$\beta_1^\star = \mathrm{argmax}_{\beta_1}\{U\}$$
$$= \frac{(n_2\beta_2 + \delta)(n_2\beta_2\delta + n_2\beta_2 r + r\delta)c_w + \sqrt{n_2\beta_2 r(n_2\beta_2\delta + n_2\beta_2 r + r\delta)(n_2\beta_2 + \delta)^3 c_w}}{(n_2\beta_2 + \delta)n_2\beta_2 r - (n_2\beta_2\delta + n_2\beta_2 r + r\delta)c_w}, \tag{50}$$

with the condition

$$c_w < \frac{n_2\beta_2 r(n_2\beta_2 + \delta)}{n_2\beta_2\delta + r(n_2\beta_2 + \delta)}. \tag{51}$$

The formula effectively relates the best response on the part of the SB to the parameters of the system, and allows analysis of the effects of each parameter on the SB. The condition (51) bounds the rate of the waiting cost such that when c_w exceeds this bound, the cost of waiting in the auction to procure the item becomes prohibitively high that the SB should perhaps seek to purchase the item directly from a seller without participating in the time-consuming auction mechanism. Of course, this may mean paying a larger sum for the item, but the solution is fitting in light of the time constraints.

In Figure 5, some results on the optimal bidding rate on the part of the SB are traced against the rate of the waiting cost c_w. The solutions obtained using the closed form expression (50), which is convenient but only gives an approximation, are compared with the exact optimal bidding rates computed numerically via the inverse parabolic interpolation algorithm [9], which yields accurate results but is computationally more demanding. Clearly the quantity $\beta_1^\star$ increases with the waiting cost rate c_w, and also with the bid rate of the others β_2. Also, the approximate solutions closely follow the exact for smaller valuations of c_w and β_2, and increasingly diverge when the quantities take higher valuations. Interestingly, it can also be observed that the approximations underestimate the exact solutions for $\beta_2 = 1$, and that the approximations overestimate the exact solutions for $\beta_2 = 0.5$ in the present setting.

Taking this further, we can generalise the solution to the first model in Section 3 which allows for an arbitrary groups of bidders with group specific bid rates. Suppose we are interested in a particular bidder's optimal behaviour with respect to the rest of

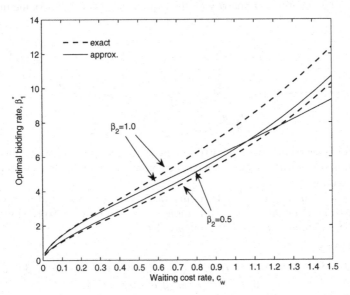

Fig. 5. The optimal bidding rate for the SB against the waiting cost rate c_w, presented for different bid rates on the part of the others β_2. Other parameters: $\delta = 0.5$, $r = 1$, $n_2 = 10$.

the participants. Let the first group consist of this single Special Bidder $n_1 = 1$, and aggregate the other bidders into groups according to bid rate that is common among members. Then the optimal bid rate for the bidder is

$$\beta_1^\star = \left[\frac{(\delta + \sum_{j=2}^{k} n_j \beta_j)}{r(\delta + \sum_{j=2}^{k} n_j \beta_j) \sum_{j=2}^{k} n_j \beta_j - (r\delta + (r+\delta) \sum_{j=2}^{k} n_j \beta_j) c_w} \right]$$

$$\cdot \left[((\delta + r) \sum_{j=2}^{k} n_j \beta_j + r\delta) c_w \right.$$

$$+ \left. \left[r(\sum_{j=2}^{k} n_j \beta_j)(\delta + \sum_{j=2}^{k} n_j \beta_j)(r\delta + (r+\delta) \sum_{j=2}^{k} n_j \beta_j) c_w \right]^{1/2} \right]. \quad (52)$$

7 Bidders with Independent and Non-identically Distributed Valuations

In this section, we deviate from the restriction that all bidders share a common valuation which may itself be represented with a random variable, and study non-identical valuations between the SB and the others. Particularly, let the valuation that the other bidders associate with the item be represented with the nonnegative integer random variable W with probability distribution $G(w) = \text{Prob}[W = w]$, while the Special Bidder has a nonnegative integer valuation V given by the probability distribution $D(v) = \text{Prob}[V = v]$. Further, assume that the seller has a minimum price $s > 0$ below which it will not sell the item. The minimum or reservation price can also be represented with a random variable, but here we take them to be constants. Thus, the case $s = 1$ and $W = V$ will reduce to the model studied in Section 4.

First consider arbitrary fixed values $V = v$ and $W = w$. The system equations for this case become:

$$P(B(2,1))((n_2 - 1)\beta_2 \sum_{w=1}^{\infty} G(w)1_{[w>1]} + \beta_1 1_{[v>1]} + \delta 1_{[s\leq 1]}) = n_2\beta_2 P(0), \qquad (53)$$

$$P(B(1,1))(n_2\beta_2 \sum_{w=1}^{\infty} G(w)1_{[w>1]} + \delta 1_{[s\leq 1]}) = \beta_1 P(0), \qquad (54)$$

$$P(B(2,l))((n_2 - 1)\beta_2 \sum_{w=1}^{\infty} G(w)1_{[w>l]} + \beta_1 1_{[v>l]} + \delta 1_{[s\leq l]})$$

$$= P(B(2,l-1))(n_2 - 1)\beta_2 \sum_{w=1}^{\infty} G(w)1_{[w>l-1]}$$

$$+ P(B(1,l-1))n_2\beta_2 \sum_{w=1}^{\infty} G(w)1_{[w>l-1]}, \quad 2 \leq l \leq v-1, \quad (55)$$

$$P(B(1,l))(n_2\beta_2 \sum_{w=1}^{\infty} G(w)1_{[w>l]} + \delta 1_{[s\leq l]}) = P(B(2,l-1))\beta_1 1_{[v>l-1]}, \quad 2 \leq l \leq v-1, \qquad (56)$$

$$P(A(2,l))r = P(B(2,l))\delta 1_{[s\leq l]}, \quad 1 \leq l \leq v, \qquad (57)$$

$$P(A(1,l))r = P(B(1,l))\delta 1_{[s\leq l]}, \quad 1 \leq l \leq v, \qquad (58)$$

$$r\sum_{l=1}^{v} [P(A(1,l)) + P(A(2,l))] = P(0)(\beta_1 + n_2\beta_2), \qquad (59)$$

$$P(0) + \sum_{l=1}^{v} [P(B(1,l)) + P(A(1,l)) + P(B(2,l)) + P(A(2,l))] = 1. \qquad (60)$$

To solve for $P(0)$, we start with rewriting (16) in the present context

$$P(0) = \frac{r}{\beta_1 + n_2\beta_2} \sum_{l=1}^{v} [P(A(1,l)) + P(A(2,l))]$$

$$= \frac{r}{\beta_1 + n_2\beta_2} \sum_{l=1}^{v} \frac{\delta 1_{[s\leq l]}}{r} [P(B(1,l)) + P(B(2,l))]$$

$$= \frac{\delta}{\beta_1 + n_2\beta_2} \sum_{l=s}^{v} [P(B(1,l)) + P(B(2,l))],$$

where the summation is adjusted for the lower limit s below which sales do not occur. Note that this differs from (17) by omitting terms $\sum_{l=1}^{s-1}[P(B(1,l)) + P(B(2,l))]$. As a consequence, it is not possible to factorise $P(0)$ out in the normalisation equation (60), which means the simple closed form for $P(0)$ (26) does not hold any longer. Nevertheless, a numerical solution can be computed for the stationary probabilities:

1. In a sequential order for each $l = 1, \ldots, v$, solve for $P(B(1,l))$ and $P(B(2,l))$, keeping $P(0)$ as the multiplying factor for each term
2. Solve $P(A(1,l))$ and $P(A(2,l))$ for all l, keeping the $P(0)$ multiplying term
3. Find $P(0)$ using the normalisation equation (60)
4. Update $P(B(1,l))$, $P(B(2,l))$, $P(A(1,l))$ and $P(A(2,l))$ for all l by multiplying $P(0)$.

In the following figures, some numerical examples are provided to illustrate the predictions of the model. To isolate and study the influence of the bidders' valuations on the outcome of the auction, we examine the case where all bidders have symmetrical bidding rates, i.e. $\beta_1 = \beta_2$. Three cases of interest are presented: one when both the distributions D and G are identically distributed, and the other two when either one distribution stochastically dominates the other.

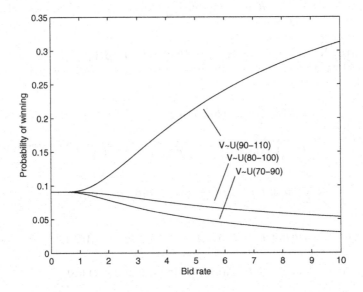

Fig. 6. The SB's winning probability $\pi(1)$ against the common bidding rate, presented for various distributions of V against the other bidders' valuation W that is uniformly distributed between 80–100. Other parameters: $\delta = 0.5$, $\beta_1 = \beta_2$, $r = 1$, $n_2 = 10$.

For the SB, having even a marginally higher valuation than the competitors, $U(90-110)$ compared with $U(80-100)$, leads to a better chance of winning the auction, but this advantage is only noticeable when the bidding rate is above some threshold (Figure 6). For bid rates below 1, the probability of winning is stagnant, because the cumulative rate of bid offers being significantly lower than the seller's decision rates leads to the auction closing at low prices. As the bid rate is increased, the competition among bidders becomes stronger and drives the closing price nearer to the limits of the bidders' valuations. Thus the SB's higher valuation becomes helpful in increasing its chances of securing a win at high bid rates. Of course this higher winning probability comes with the cost of higher price paid upon winning. Following the same logic, when the SB has a lower valuation than the competitors, i.e. $U(70-90)$, it is disadvantaged as the game gets more competitive at high bidding rates, and has less chances of winning. Perhaps the more surprising and interesting result is the case when the SB's valuation is distributed identically to the others $V \sim U(80-100)$, for which the probability of winning decreases with higher bid rate, rather than remaining unchanged. In the model, although the valuations differ between the SB and the others, within the group of the other bidders, the valuations are taken to be identical; it should be noted that this is a stronger property than the valuations being identically distributed. For illustration purposes, consider the simplifying case when one's valuation can only take two discrete outcomes: high or low. In any instantiations for which the SB and the others have valuations that are both high or both low, the outcome of the auction to the SB in terms of the winning probability will not change. However, the gain that the SB enjoys when its valuation is high and the others' valuation is low is outweighed by the loss in the winning probability when its valuation is low and the others' high, simply because there are n_2 others who identically share high valuations and accentuate the difference. Hence we can expect the situation to be exacerbated when there are more of the other bidders present, and the gradient of the drop in winning probability to be steeper with the bid rate.

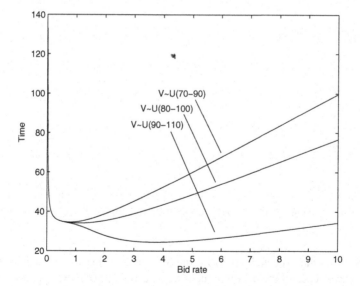

Fig. 7. The SB's expected time to win ψ against the common bidding rate, presented for various distributions of V against the other bidders' valuation W that is uniformly distributed between 80–100. Other parameters: $\delta = 0.5$, $\beta_1 = \beta_2$, $r = 1$, $n_2 = 10$.

The quantity ψ or the SB's expected time to win is traced in Figure 7 against the bid rate on the x–axis. Interestingly, the waiting time drops exponentially when the bid rate is increased from a very small nonzero value to about 1. For the case where the SB has the same valuation or a valuation that is smaller on average than the others, the time to win increases for bid rates above this threshold, and this increase is sharper for the latter case. This observation is to the contrary of the previous result in Figure 2, in which the quantity ψ is strictly decreasing with β. Because now all $n_2 + 1$ bidders share a common bid rate, increasing the rates beyond what is minimally necessary to secure a reasonable chance of winning will not reduce the time to win for the SB indefinitely; hence it will not be strictly decreasing. We should then inquire as to why the quantity increases at higher bid rates rather than staying constant, since all bidders share the same rate. The reason ties in with the smaller winning probability for the SB at higher bid rates, as illustrated in Figure 6. Incidentally, from the same figure, the case for SB having a higher valuation than the others on average ($V \sim U(90–110)$) has the unique property of recording higher winning probabilities at high bid rates, and this is reflected in the smaller waiting time to win in Figure 7.

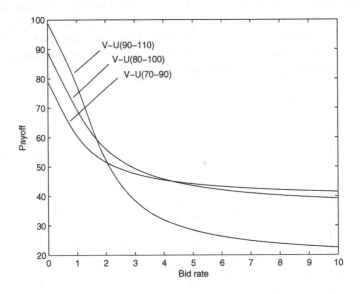

Fig. 8. The SB's expected payoff ϕ against the bid rate for various distributions of V against the other bidders' valuation W that is uniformly distributed between 80–100. Other parameters: $\delta = 0.5$, $\beta_1 = \beta_2$, $r = 1$, $n_2 = 10$.

Figure 8 shows the expected payoff for the SB when it wins an item against the bid rate on the x–axis. Clearly the payoff reduces with the bid rate. At low bid rates, the case for $V \sim U(90–110)$ records the highest payoff, followed by the other cases in descending order of the average valuation, as might be intuitively expected. Intriguingly, as the bid rate increases the curves criss-cross each other. Comparing the case for $V \sim U(90–110)$ with $V \sim U(70–90)$, it is observed that beyond the bid rate of 2 the SB enjoys a higher payoff in the latter case than the first. In other words, for bid rate above the valuation 2, the SB would receive a higher payoff if it had a smaller valuation on average than if it had had a higher valuation on average than the others. This may seem somewhat counter-intuitive at first, but revisiting the definition (35), it should be recalled that the quantity is the expected savings made with respect to the maximum that the SB is willing to pay *conditional on the outcome of the auction that the SB actually wins*. With $V \sim U(90-110)$, because the other bidders will gradually fall out of competition as the price increases beyond 80 and totally withdraw when the price reaches 100, a sale is more likely to be made to the SB than the others at these prices. These high prices contribute to the drop in the payoff, but it must be noted that as a direct consequence of being more likely to win at high prices the SB enjoys a shorter waiting time to win (Figure 7).

8   Concluding Remarks

Economic activities over the Internet with worldwide connectivity or over specialised islands of computer networks have become more common with advancements in computer communications technology and decreasing computational costs. Humans are increasingly supported by or in some cases replaced with software agents that are specifically designed for interacting and negotiating on their behalf. With the benefits of time-saving aspects, real-time monitoring, and higher computational accuracy leading to better rule-based decisions of this automation, it is important that these activities are studied in-depth.

In this chapter, we develop models for studying auctions as economic mechanisms that have well-defined protocols and rules to ensure timely conclusions and principled outcomes. To this end, we employ tools and techniques from stochastic modelling theory. We show how these techniques can be used to build auction models of increasing complexity using simple first principles. In the models, the auction proceeds in an ascending price order and the process is represented as a jump process that takes valuation from a discrete state-space which contains information on the identity of the leading bidder and the price that it has offered for the item on sale. Moreover, auctions are infinitely repeating as independent and yet probabilistically identical to each other.

We start the chapter by formalising the problem in a general form, allowing for an arbitrary number of groups of bidders to participate in an auction where each group is characterised with some common behaviour among the members. The variables that dictate the evolution of the process are the bid increment offered by a bidder over the previous and the time taken to place the offer, the maximum valuation that the bidders associate with the item, the decision time taken by the seller to consider the offer, and the speed with which the auction restarts after each cycle. By restricting our attention to the case where the bid increments are always one, and the bidding times and the seller's decision times are exponentially distributed random variables, we propose a Markov process to model the problem. With this the probabilities of each outcome of the auction, in terms of which bidder wins and the sale price, can be evaluated. Due to its general form, the model does not admit analytical solutions, but we outline how the probabilities can be computed numerically.

Following that, we turn our attention to the case of a single Special Bidder (SB) who plays against the other participants in the auction, where, for tractability, all other bidders are aggregated into one group with a common bid rate. We show how this reduced model admits analytical solutions for the probabilities of interest, unlike the previous. The measures of interest that determine the SB's performance in the auctions are then introduced: the average time that it waits until it wins the item and the average savings that it makes in the cost of purchase with respect to the maximum price that it is willing to pay. Via numerical examples, we show that these two quantities both decrease with increasing bid rate, hence there is a trade-off between buying a good quickly and the high cost that is paid for it.

The measures of interests for the SB can be computed with sequential summations of terms involving the state probabilities. However, in pursuit of mathematical tractability, we introduce an approximate model with which the measures can be expressed in closed form. Consequently, we also able to solve an optimisation problem that balances the trade-off the SB faces, and obtain an analytical expression that approximates the optimal bid rate for the SB. In the final part, we study an extension of the model in which the SB may have a valuation that is distributed differently from the other bidders' valuations, and illustrate the its predictions with numerical examples.

The primary contribution of this chapter is in illustrating how probabilistic modelling techniques that are popular among the operations research community can be used to study automated auctions. The approach can be easily extended to analyse more elaborate systems. Also, we have assumed perfect knowledge on the part of the bidder with regard to the parameters of the system, and proceeded to predict outcomes relying on this knowledge. However the problem of learning or estimating the parameters from observations, which is a practical issue, is not straightforward and should be explored. Yet another interesting question is the case when the valuation of the item is common to all, but every bidder has only partial information that provides some estimate of the valuation.

Acknowledgements. The work presented in this chapter was undertaken as part of the ALADDIN (Autonomous Learning Agents for Decentralised Data and Information Networks) project and is jointly funded by a BAE Systems and EPSRC (Engineering and Physical Research Council) strategic partnership (EP/C548051/1).

The Authors

Erol Gelenbe is the Professor in the Dennis Gabor Chair and Head of the Intelligent Systems and Networks Group at Imperial College London. His current research interests include the performance of computer systems and networks, and probabilistic models of computations and systems. Recent research results include autonomic Quality of Service based routing in self-aware networks. A native of Istanbul and graduate of the Middle East Technical University in Ankara, Turkey, he is a Fellow of ACM, IEEE and IET, and a member of Academia Europaea, Hungarian Academy of Sciences, Turkish Academy of Sciences, and French National Academy of Engineering. He has some 120 papers in the major journals of computer science and electrical engineering, as well as in journals of applied probability and operations research. He has received honorary doctorates from University of Rome Tor Vergata (Italy), Bogazici University (Istanbul) and Liege University (Belgium), and scientific prizes both in Turkey and France. He is a Commander of the Order of Merit of Italy, and an Officer in the Order of Merit of France.

Kumaara Velan is a research assistant in the Intelligent Systems and Networks Group at the Department of Electrical and Electronic Engineering at Imperial College London, where he is working towards a PhD. His main interests include probability models, performance analysis and simulation of computer systems, auctions, and economic mechanisms. He received his M.Sc. (Eng.) with distinction in data communications from The University of Sheffield.

References

1. Gagliano, R.A., Fraser, M.D., Schaefer, M.E.: Auction allocation of computing resources. Commun. ACM 38(6), 88–102 (1995)
2. Gelenbe, E., Mitrani, I.: Analysis and Synthesis of Computer Systems. Academic Press, New York and London (1980)
3. Gelenbe, E.: Analysis of Automated Auctions. In: Levi, A., Savaş, E., Yenigün, H., Balcısoy, S., Saygın, Y. (eds.) ISCIS 2006. LNCS, vol. 4263, pp. 1–12. Springer, Heidelberg (2006)
4. Gelenbe, E.: Analysis of Single and Networked Auctions. ACM Trans. Internet Technol. 9(2), 1–24 (2009)
5. Gelenbe, E., Györfi, L.: Performance of Auctions and Sealed Bids. In: Bradley, J.T. (ed.) EPEW 2009. LNCS, vol. 5652, pp. 30–43. Springer, Heidelberg (2009)
6. Gelenbe, E., Velan, K.: An Approximate Model for Bidders in Sequential Automated Auctions. In: Håkansson, A., Nguyen, N.T., Hartung, R.L., Howlett, R.J., Jain, L.C. (eds.) KES-AMSTA 2009. LNCS, vol. 5559, pp. 70–79. Springer, Heidelberg (2009)
7. Guo, X.: An optimal strategy for sellers in an online auction. ACM Trans. Internet Technology 2(1), 1–13 (2002)
8. Maes, P., Guttman, R.H., Moukas, A.G.: Agents that buy and sell. Commun. ACM 42(3), 81–91 (1999)
9. Press, W.H., Teukolsky, S.A., Vetterling, W.T., Flannery, B.P.: Numerical Recipes in C++: The Art of Scientific Computing, 2nd edn. Cambridge University Press (2002)
10. Rothkopf, M.H., Harstad, R.M.: Modeling Competitive Bidding: A Critical Essay. Manage. Sci. 40(3), 364–384 (1994)
11. Rothkopf, M.H.: Decision analysis: The right tool for auctions. Decision Analysis 4(3), 167–172 (2007)
12. Velan, K.: Modelling Bidders in Sequential Automated Auctions. The Computer Journal 53(2), 208–218 (2010)
13. Velan, K., Gelenbe, E.: Analysing Bidder Performance in Randomised and Fixed-Deadline Automated Auctions. In: Jędrzejowicz, P., Nguyen, N.T., Howlet, R.J., Jain, L.C. (eds.) KES-AMSTA 2010, Part II. LNCS, vol. 6071, pp. 42–51. Springer, Heidelberg (2010)
14. Zeithammer, R.: Forward-Looking Bidding in Online Auctions. J. Marketing Res. 43(3), 462–476 (2006)

Generating B2C Recommendations Using a Fully Decentralized Architecture

Domenico Rosaci and Giuseppe M.L. Sarné

DIMET, Università Mediterranea di Reggio Calabria
Via Graziella, Località Feo di Vito
89122 Reggio Calabria, Italy
{domenico.rosaci,sarne}@unirc.it

Abstract. In the last years, Business-to-Consumer (B2C) E-Commerce is playing a key role in the Web. In this scenario, recommender systems appear as a promising solution for both merchants and customers. However, in this context, the low scalability of the performances and the dependence on a centralized platform are two key problems to face. In this paper, we present a novel recommender system based on a multi-agent architecture, called Trader REcommender Systems (TRES). In TRES, the agents exploit their user's profiles in their interaction, to make the merchants capable to generate effective and efficient recommendations. The architecture we have adopted is fully decentralized, giving to each merchant the capability to generate recommendations without requiring the help of any centralized computational unit. This characteristic, on the one hand, makes the system scalable with respect to the size of the users' community. On the other hand, the privacy of each customer is preserved, since the merchant retrieves information about each customer simply monitoring the customer behaviour in visiting his site.To show the advantages introduced by the proposed approach some experimental results carried out by exploiting a prototype implemented in the JADE framework are presented.

1 Introduction

In these years, E-Commerce (EC) activities are playing a key role in the Web as attested by the increasing number of commercial transactions therein performed. As a consequence, a significant number of powerful and sophisticate tools have been recently developed for supporting traders in all their commercial processes with a high automation level. In particular, a great attention has been reserved to the Business-to-Consumer (B2C) activities, comparable with the retail trade of traditional commerce, both from the customers that can exploit the opportunity offered by the EC for a simple and comfortable access to an open-world market without time and space boundaries, and from the merchants that can offer their products to a wide audience by using a convenient media [1, 2].

A. Håkansson & R. Hartung (Eds.): *Agent & Multi-Agent Syst. in Distrib. Syst.*, SCI 462, pp. 163–185.
DOI: 10.1007/978-3-642-35208-9_9   © Springer-Verlag Berlin Heidelberg 2013

1.1 Motivations: Recommender Systems can Effectively Support EC

In the aforementioned scenario, recommender systems [3–7] are among the most meaningful applications both for merchants and for customers. In particular, the merchants can take advantage of the improvement of the performances for their e-Commerce sites; on the other hand, the customers are supported in their decision-making process by means of some useful suggestions about objects, products, or services potentially interesting for them. Different techniques have been proposed in the literature to implement recommender systems in order to generate effective suggestions. Based on the adopted technique, recommender systems are generally partitioned in three main categories [5], namely: (*i*) *Content-based*, that suggest to a user items which appear the most similar to those he has already accessed in the past; (*ii*) *Collaborative Filtering*, that suggest to a user items which have been also considered by similar users; (*iii*) *Hybrid*, that combine both content-based and collaborative filtering techniques to generate recommendations. This latter approach is usually recognized as the most promising solution [5].

To carry out such an activity, recommender systems need to exploit a suitable representation (*profile*) of the customer's interests and preferences. This representation is automatically derived by monitoring and interpreting the large amount of information that each user spreads during his/her Web trading activities. These data can be implicit data as purchase histories, Web logs, cookies, etc. and/or explicit data that a user can provide through his/her beaviour. In particular, in the automatic construction of such a profile the most part of the existing recommender systems mainly consider only the number of accesses to a specific product or to a product category. In the EC field, and specifically in B2C activities, this approach risks to be misleading because a trading activity consists of more different phases and usually the interest degree corresponding to two accesses to a product in two distinct trading phases (e.g., a visit and a purchase) is different.

1.2 Our Contributions: A Behavioural Model to Represent B2C Processes and an Agent-Based Recommender System for EC

In order to realize a faithful representation of the real customer's interests and preferences and consequently to generate more effective suggestions, in this paper we propose a recommender system, called *Trader REcommender System* (TRES). This recommender system exploits users' profiles that, preserving customer's privacy (see Section 3), take into account their interests and preferences in the different phases involved by a Web B2C process. In such a manner, it will be possible to suitably consider in the interest computation each access to a product with respect to its real trading context.

Then, the first contribution of our paper consists of a model to represent business-to-customer activities in the Web context. In the past, other behavioural models have been introduced to capture the different phases carried out by enacting an EC process. Several of such models are extensions of other ones designed for the conventional commerce, such as the Nicosia Model [8] or the Engel and Blackwell Model [9], or explicitly thought for EC, such as the Nissen's Commerce Model [10] or the E-commerce Value Chain Model [11]. Another widely known approach is the Consumer Buying Behaviour (CBB) model [12], structured in six different phases, namely: (i) "Need Identification" - a user identifies his/her needs to satisfy; (ii) "Product Brokering" - a potential customer searches for products able to satisfy his/her needs; (iii) "Merchant Brokering" - when it is known what to purchase a consumer searches a merchant from whom to buy the chosen goods or services; (iv) "Negotiation" - transaction terms (i.e. price, quantity, quality of service, etc.) are fixed in this phase; (v) "Purchase and Delivery" - a customer fulfils the purchase choosing a payment option and a delivery modality, among those available; (vi) "Service and Evaluation" - a customer can estimate his/her satisfaction degree about a purchase. For the CBB model many and different extensions have been proposed, as for example the $E - CBB$ [13] and the $E^2 - CBB$ [14] to consider emerging behaviours as the formation of coalitions and the EC-site visits, respectively.

However, we remark that some modifications can be made to these models in order to more effectively represent EC activities carried out over the Web. In particular, with respect to the original CBB model, we observe that several trading processes allow the first three CBB stages to be unified without any substantial difference. In fact, it is not possible for all the B2C processes to effectively split the customer's behaviour in these three phases (e.g., they could be performed almost entirely at the same time during a B2C process over a Web EC site, where he/she can contextually discovery a need, find an item and find a merchant able to satisfy it). For such a reason, we think that the choice to unify the first three CBB stages can improve the efficiency of a behavioural model without significant losses of precision; thus we consider as a unique activity (i.e. stage) all the actions performed by a user from when he/she understands to have a need to when he/she knows how to satisfy it. Moreover, often the negotiation stage currently involves also the delivery mode, differently from the original CBB model that only considers the price negotiation of the desired product. Consequently, the purchase phase only consists of the purchase order that represents the effective customer's will to perform that purchase. Finally, in the adopted behavioural model the last CBB stage is not considered because it is implicitly considered in the customer's profile; in fact only a satisfied consumer will perform in the future other similar purchases.

We argue that these characteristics allow us to more effectively model the actual traders' activities in only three phases, respectively called "Visit", "Negotiation" and "Purchase", than the others proposals present in the literature. In the following of this paper we refer to this Web Trader Model (WTM) to represent the B2C processes of a customer. A graphical comparison between WCM and the CBB model is depicted in Figure 1

A promising solution to implement recommender systems that have to deal with different Web sites is represented by the agent technology in which *software information agents* [15–22]) autonomously and proactively perform some tasks

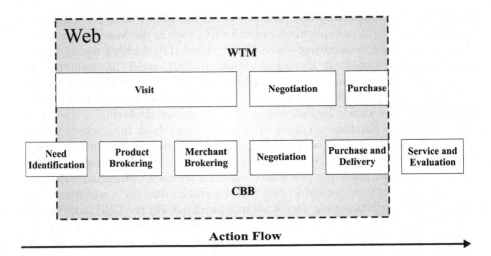

Fig. 1. Comparison between the Web trader model (WTM) and the Consumer Buying Behaviour model (CBB)

on the behalf of its human user. Just as a second contribution of our paper, we propose an approach based on a multi-agent architecture to recommend EC activities. Specifically, in the proposed context each trader is associated with a software agent that constructs and stores an internal profile to take into account his/her whole Web trading history. Furthermore, an agent should adopt suitable techniques to generate an initial user's profile for beginning to work. Note that to allow the updating of the traders' profile, with respect to all the visited EC sites, the profiles can not be stored in the Web sites while the traders navigate on different sites. The agents exploit their user's profiles in the interaction with the EC sites in order to generate effective recommendations based on both content-based and collaborative filtering techniques for adapting the site presentation with useful suggestions. The purposes of our system are (*i*) to produce better suggestions than the traditional approaches, due to the adoption of a suitable customers' behaviours representation and (*ii*) to show high efficiency in the construction of users' profiles, due to the distributed multi-agent architecture.

1.3   Evaluation of the Proposal

To evaluate the effectiveness of TRES for generating suitable suggestions, we have implemented a message-based agent platform to support customers and merchants during all the phases of a B2C process, arranged accordingly to the WTM. In the TRES platform each trader is associated with a personal XML-based agent that monitors his/her B2C activities and specifically (*i*) a customer's agent automatically builds a profile of its user by monitoring his/her interests and preferences during all his/her EC site visits, while (*ii*) a merchant's agent, associated with an EC site (i.e., a merchant), automatically builds a profile by collecting interests and preferences shown by the visitors of its site. TRES agents exploit their profiles to take care of the traders in a personalized and homogeneous way by means of reciprocal message-based interactions. As a result, customers will be provided during their EC-site visits with personalized Web presentations built on-fly exploiting the suggestions generated by using both a content-based and a collaborative filtering approach. The other advantages of the TRES platform can be summarized as follows:

- The usage of the Extensible Markup Language (XML) [23] allows: *(i)* to unify the representation of products belonging to various categories and catalogues overcoming several heterogeneity problems (platforms, languages, applications and communication modalities); *(ii)* to manage the formalization of agent messages and profiles in a light and easy way; *(iii)* to formalize the agent communications in the versatile Agent Communication Markup Language (ACML) [24], an XML coding of the FIPA-ACL [25, 26], that offers some significant advantages as the usage of existing XML message parsers and the opportunity to enrich messages with a large variety of features (i.e., SSL, links, etc.)
- A customer, in order to preserve his/her privacy, can decide which information to send to a merchant agent for building on-the-fly a personalized Web site presentation.
- EC actors can exploit all the TRES features described above in an easy way by means of their usual Web-browsers.

We have carried out some experiments, over the TRES platform, to evaluate the performances of TRES in supporting B2C traders. The experiments have confirmed our expectations and the performances shown by TRES show significant improvements if compared to other similar profile-based recommender systems [27, 28].

The paper is organized as follows: in Section 2 we present the TRES framework and the activities performed by the agents in order to monitor customers' behaviours. Section 3 describes the TRES algorithm while Section 4 presents the Related Work. In Section 5 some experiments are presented and TRES performances are discussed. Finally, in Section 6, we draw some final conclusions.

2   An Overview of the TRES Framework

In this section, the TRES framework will be described in detail. In such a framework (see Figure 2) each customer C (resp. merchant M) is associated to a personal agent c (resp. m) logged into the TRES Agency (Ag). Below we will describe the knowledge representation model used by the agents (Section 2.1), the structure of the agents (Section 2.2) and the agency (Section 2.3), as well as the trading support (Section 2.4) provided by the agents.

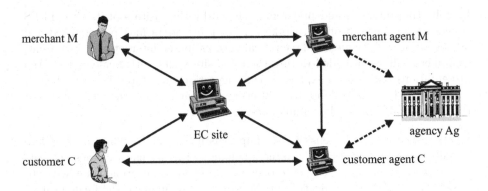

Fig. 2. The TRES framework

2.1 Representation of Objects and Categories of Interest

TRES agents support traders in their commercial tasks for products that meet their interests. To this aim, all the TRES agents share the same *Catalogue* $\mathcal{C}$, representing the common agent knowledge. In this catalogue each *product category* (*pc*) of interest is described with a pair (*code*, *td*), where *code* is the identifier of the product category and *td* is a *textual description*.

Usually, each Web page includes more product instances belonging to different product categories. It is reasonable to assume that each time that a user clicks on a hyperlink pointing to a product instance *pi* (and its associated product category *pc*), this means he/she shows an *interest* for *pi* (resp., *pc*). These interests are measurable by taking into account frequency and context of their access. The TRES agents build their users' profiles by monitoring all the product instances and product categories accessed by the customers in order to measure the associated interests. Moreover, to determine collaborative filtering recommendations, each merchant agent also computes the similarity among its visitors by exploiting the values of interest shown in the offered product instances (see below).

Currently, the TRES catalogue adopts the six digit edition of the *NAICS* coding [29], an official, public, hierarchical classification used in North America to classify businesses in categories. The catalogue is implemented by means of a simple XML-Schema [30] where a product category is described by using the notion of *element* and the product instances are represented by XML *element instances*.

2.2 The Agents

Each generic user U (i.e. a customer C or a merchant M) is associated to a personal agent a (i.e. a customer agent c or a merchant agent m), which manages the *User Profile* (*UP*) of U, graphically described in Fig. 3. In particular, this user profile consists of a tuple $\langle UD, WD, BD \rangle$, where:

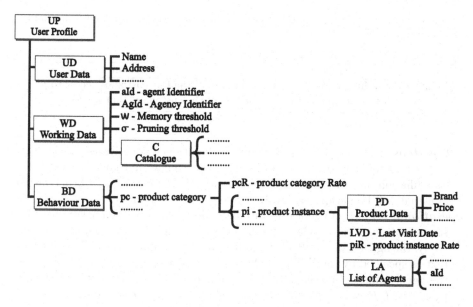

Fig. 3. The User Profile (UP)

- the *User Data (UD)* stores personal U's data as name, address, financial data, etc.
- the *Working Data (WD)* collects the identifiers of the agent (aId) and the agency $(AgId)$, the *Memory* parameter (ω) used in the computation of the concept and instance rate, the *Pruning* threshold (σ), described below, and the current product catalogue $\mathcal{C}$.
- the *Behaviour Data (BD)* of a user contains some parameters that describe the past behaviour of the user. Moreover, even though customer and merchant agent share the same structure of profile, some parameters have different meaning for the two types of agent (see Table 1). More in detail, if U is a customer (resp. merchant), BD stores information on each product instance pi and corresponding category pc accessed by the customer (resp. offered by the merchant). Specifically, BD consists in a list of product categories accessed/offered by U and for each category the profile stores its rate pcR (see below) and a list of product instances visited/offered by U and belonging to that category. Furthermore, for each product instance in BD, the profile stores (in PD) some information associated to it (e.g., brand, price and so on), the last visit (LDV) to that instance, its rate piR (see below) and a list of agent (LA) related to that instances.

The behaviour of an agent can be summarized in only two main steps identified as: *(i)* **setup**, that includes some simple semi-automatic procedures to affiliate/delete an agent a to/from the TRES framework; *(ii)* **operational**, that consists of all the tasks to support trading activities (note that the recommendation generation will be described in Section 3). More in detail:

Table 1. Parameters stored in BD having a different meaning for a customer and a merchant agent

Name	Customer	Merchant
LVD	Date of his/her last visit to pi	Date of the last customer access to pi
L	List of the merchants that sell pi	List of the customers that accessed to pi

setup steps: they are performed *(i)* when the agent a is activated, some basic UP parameters have to be set and sent to the agency Ag that will provide for its affiliation into the agent community and *(ii)* in presence of a deactivation of a that will be performed by Ag for a specific U's request.

operational steps: an agent a is automatically activated when a Web activity starts and deactivated when it ends or for an explicit user's choice. Operatively, the agent a monitors each trading activity, taking into account its context as specified by the WTM model. After each access to a product instance, the agent updates/sets its $BD \in UP$. Furthermore, each agent periodically prunes its UP from negligible data based on both the pcR and σ values.

In other words, when a customer visits an EC site associated with a merchant, both their agents monitor the customer's behaviour in his/her visit (i.e., the accessed product instances and the trading context, accordingly to the WTM). We remark that TRES is a message-based framework in which a message is exchanged between agents due to *(i)* a trader's decision (i.e., a product research) or *(ii)* an action performed by its owner (i.e., a purchase) or *(iii)* an automatic running process (i.e., during an automatic negotiation). In this way, agents easily monitor all the trading activities by means of the exchanged messages, while customers and merchants can access to a message content by using a Web interface. Then, as a consequence of such a monitoring activity, a customer (resp., merchant) agent, with respect to a product instance pi belonging to a product category pc, could take into account the interest rate piR and pcR for that pi and respective pc. More in detail, we propose to compute piR and pcR by means of the following formulation:

$$piR = (1-\omega) \cdot \frac{piR}{log_{10}(10+\Delta)} + \omega \cdot \frac{\rho_i}{log_{10}(10+N_i)}$$
$$pcR = (1-\omega) \cdot \frac{pcR}{log_{10}(10+\Delta)} + \omega \cdot \frac{\rho_i}{log_{10}(10+N_i)}$$

where $\omega \in [0;1]$, is a system parameter that represents the "memory" of piR and pcR with respect the past trading activities of the customer. High values for the ω parameter give more relevance to the most recent accesses. Moreover, Δ is the temporal distance expressed in days between the date of the current visit and LVD, and ρ is the weight assigned to the performed WTM stage. We have assigned for the "Visit", "Negotiation" and "Purchase" stages the following values, based on the monitored customers'behaviour (see Section 5): $\rho = 0,01$, $\rho = 0,25$ and $\rho = 1$. Finally, N_i is the number of the same stages performed for the same product in the same day.

Therefore, the parameters piR and pcR, by means of ω, are able to take into account the whole history of accesses to a product (instance and category) suitably weighting both the past and the current accesses. Furthermore, the past customer's interest in a product is updated to the current date based on the time (expressed in days) passed from his/her last access to it. The current WTM stage is considered by exploiting the ρ_i values weighted by $1 + log_{10}(N_i)$, where N_i is the number of times the i-th WTM stage has been performed in the same day. In this way, each time a WTM stage is repeated in the same day, its contribution is considered in a decreasing way because it derives by the same customer's interest.

To illustrate how piR (pcR) works, we present two examples, where we set $\omega = 0.1$. As for the first example, we consider a customer that searches for a new product and carries out the "Visit" WTM phase ($\rho = 0, 01$), thus LVD will be set to the current date, while piR (pcR) is equal to 0.001. If the customer decides to immediately performs also the WTM "Negotiation" phase ($\rho = 0, 25$), then LVD will be set to the current date and piR (pcR) will be 0.0259. If the customer performs also the purchase of the product, then LVD does not change and the value of piR (pcR) will be equal to $0, 1233$. As for the second example, consider the case of a product accessed in the WTM 'Visit" phase twenty days ago. If today the customer accesses three consecutive times for the same WTM phase, then piR (pcR) will assume after each access the values $0, 0016$, $0, 0022$ and $0, 0027$, respectively.

2.3   The Agency

The Agency manages (in terms of insertions, deletions and updating) its Agency Profile (AgP), describes agents affiliations/deletions and provides them with some services.

The AgP (Figure 4) includes: *(i)* the *Agency Identifier* ($AgId$); *(ii)* the *System Data* (SD) that contains the *Agent Pruning* threshold (Σ) exploited to deallocate long-time inactive agents, the *Memory* (ω) and the *Pruning* (σ) thresholds, that have been already described in Section 2.2, and the current catalogue $\mathcal{C}$; *(iii)* the *agent List* (aL), where we store the identifiers of all the affiliated agents. The behaviour of the Agency, except for some activities presented in Section 2.4, can be described as a two-steps process, namely:

affiliate managing: *Ag* automatically carries out the following operations: *(i)* When *Ag* receives an *Affiliation Request* it replies by sending an agent (aId), the system parameters ω and σ and the current catalogue $\mathcal{C}$. At this point the agent is logged in and becomes active in the TRES platform; *(ii)* *Ag* deletes an affiliate following a specific user's request or in an automatic way after an inactivity time greater than the pruning threshold Σ (in such a way the potential growth of inactive affiliates is limited).

service managing: *Ag* provides the agent community with a yellow page service.

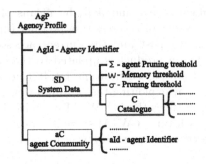

Fig. 4. The Agency Profile (AgP)

2.4   The Trading Support

TRES is a message-based framework designed to transfer, in a consistent and efficient way, business information in order to permit supporting and monitoring the activities carried out in a B2C process, accordingly with the WTM model, by merchant and customer agents. The different activities occurring within a B2C process require the exchange of several message typologies that will be briefly described below together with the different types of agent interactions. Note that in the following notations the first subscript always identifies the sender and the second the receiver, while *data* is an XML document, whose content is context sensitive and it is structured in three sections: *(i) Header*, containing some information like to sender, receiver, WTM stage involved, etc.; *(ii) Products*, that encodes some product data; *(iii) Financial*, that is relative to all the financial information needed to perform a payment. More in detail, the TRES messages are denoted as follows:

- $INF_{x,y}(data)$: it requires/provides commercial information about a product;
- $REQ\_INV_{c,m}(data)$: it requires an invoice for a product offered by M;
- $INV_{m,c}(data)$: it contains the invoice required by $REQ\_INV_{c,m}(data)$;
- $PP_{x,y}(data)$: it is used to negotiate any commercial detail not fixed or specified (i.e., the price for product without fixed price);
- $PO_{c,m}(data)$ (resp., $PO\_A_{m,c}(data)$, $PO\_R_{m,c}(data)$): it is the purchase order relative to $INV_{m,c}(data)$ (resp., the notify if it is accepted or refused);
- $MTO_{c,m}(data)$ (resp., $MTO\_A_{m,c}(data)$, $MTO\_R_{m,c}(data)$): it notifies that the payment has been performed (resp., accepted or refused) w.r.t. $PO_{c,m}(data)$;

As previously remarked, the interactions occurring between two trader agents (graphically represented in Figure 5) are performed by exchanging messages as it is below syntectically described for each WTM stage. Note that, payments are a relevant issue within any trading event and this is particularly true in an EC scenario where the presence of a network introduces some critical issues [31, 32] absent in traditional payment means. However, in this paper we do not face here this matter because for our aims it is an orthogonal question. In the following, we assume that when a purchase is performed then a payment has to occur without specifying any detail.

More specifically, the agent interactions occurring in TRES in support of customers and merchants can be summarized as:

- **Visit stage (s=1).** During an EC-site visit a customer agent can receive information about a product proposed from a merchant by means of a message $INF_{m,c}$. This message can be explicitly required by the customer to the merchant with a message $INF_{c,m}$ or implicitly when him/her clicks on a hyperlink of the EC-site to a product.
- **Negotiation stage (s=2).** In this stage all the trading details are set, as price, quantity, delivery modality and so on. More in detail, this stage starts with the messages $REQ\_INV_{c,m}$ and $INV_{m,c}$ exchanged between the traders. If the customer accepts the customer's commercial proposal contained in $INV_{m,c}$, then this stage ends. Otherwise, if it is possible, the two traders use in an alternate manner $PP_{x,y}$ messages to negotiate all the trading details; when an agreement has been risen to fix it then the $REQ\_INV_{c,m}$ and $INV_{m,c}$ messages will be again exchanged.
- **Purchase stage (s=3).** Finally, after a "Negotiation" stage, if the customer wants to purchase a products he/she has to sent the message $PO_{c,m}$ to the merchant that can accept/refuse the business by means of $PO\_A_{m,c}(data)/$ $PO\_R_{m,c}(data)$. If the purchase order is accepted, then the customer performs the payment and informs the merchant with an $MTO_{c,m}$ message. He/sheshould receive an $MTO\_A_{m,c}$ or $MTO\_R_{m,c}$ message to confirm the reception.

3 The Recommendation Algorithm

This section presents the recommendation algorithm exploited by TRES to generate suggestions based on both a *content-based* and a *collaborative filtering* approach. A computational complexity analysis of the proposed algorithm ends the section.

The algorithm has been conceived to generate suggestions by exploiting the behavioural information stored into the agent profile of the merchant and of the customer agents. Remember that the information stored into the merchant agent profile are relative to the customers that visited the associated EC site in the past; while the customer agent profile stores only information relative to its owner. In order to produce consistent recommendations, all the piR and pcR values stored into the two agent profiles are updated before of their use to the current date by means of the coefficient $log_{10}(10 + \Delta)^{-1}$, where Δ has the meaning already explained in Section 2.2. As a result, the product instances supposed the most interesting for the customer, because they match customer's interests and preferences and/or those of other similar customers, are exploited to build on-fly personalized presentation of the visited EC site. In such a way, it is possible to support a customer during its B2C process and, consequently, to improve the commercial performances of the EC site.

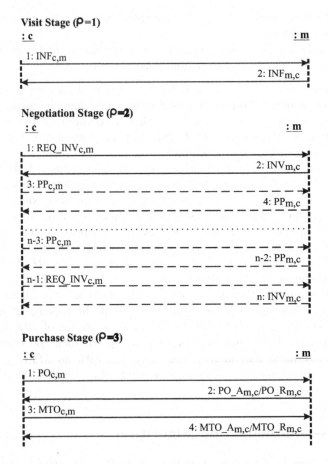

Fig. 5. The TRES support performed for the W-CBB stages

3.1 The Algorithm

The recommendation algorithm of TRES, represented by the function
`Recommendations` (Figure 6), is performed by the merchant agent (m)
when a customer agent (c) visits the m-monitored EC site. The input of
`Recommendations` is a customer agent c and its output are the lists $L3$ and $L4$ of
product instances exploited by m to build on-fly personalized Web site presentation for c. Within this function, another function `extract_pc` is called; it receives
as input the BD section of the m profile and returns the list $L1$ containing those
product categories belonging to C and currently sold by the merchant. Then the
list $L1$ is sent to c by using the function `send`, while the function `receive` waits
for the response of c consisting of the list $L2$. This list includes the v product
categories that better meet the interests of the customer (where v is a parameter
arbitrarily chosen by c). When $L2$ is received, the function `contentbased_pi` is
called and returns the list $L3$ containing the first y product instances having the

highest rate for each one of the v product categories stored in $L2$ (also y is a parameter arbitrarily set by m).

The next step deals with the construction of the array of lists, called PC, where each array element is a list associated to an agent c monitored by m and contains the v most interesting product categories. To this aim the function customersInterests is called and it receives as input the BD data section of the profile of m and the integer v. This is the most expensive function of the recommendation algorithm. Each array element $PC[\ ]$ is a list constructed (i) by computing for that agent the sum of the piR values of all the product instances for each product category that have met an interest in the past (each sum represents a global measure of the agent interest in a product category) and (ii) by ordering such sums in a decreasing order and selecting, for each agent, the v product categories having the highest sum value.

When PC has been computed, the function collaborativefiltering_pi is called; it receives as input the list $L2$ (provided by c), the array of lists PC, the BD section of the merchant agent profile and the two integers z and x, arbitrarily set by m. This function exploits PC to compute the similarity degree between c and the agents that have interacted with m in the past; this is a measure used by m to select the z agents most similar to c. Thus, for each product category in $L2$ considered by an agent, all the product instances having the most high piR values are selected and inserted in the list $L4$. $L4$ denotes the function output such that, for each of the z selected agents it contains the first x product instances with the highest rate. More in detail, the similarity is computed as the sum of the contributions provided by each pair of elements common to the two lists. For example, suppose that the product category pc_j belongs both to $L2$ and $PC[k]$, this latter associated with the generic agent k; its contribution to the similarity degree between c and k is given by $(v-|pos(pc_j^{L2})-pos(PC[k]_j)|)*(v-pos(pc_j^{L2}))$, where v is the number of product categories inserted by c in $L2$ and pos is an integer that identifies the ordinal position of the $j-th$ product category pc in a list (i.e., 1 for the first element, 2 for the second element and so on). Note that in the content-based phase, m does not look at the information about c stored in its profile because they are considered less informative than those contained in $L2$.

On the customer agent side, when the list $L1$ coming from the merchant agent m is received, the function productOfInterest is executed. In productOfInterest, the function extract_pc is called (the same used by m) to obtain the list $L5$ containing the product categories of interest for C. After that, the function intersection_pc is called and the intersection $L1 \cap L5$ is computed. The function select_pc receives as input the list $L6$ and an integer v and then it orders $L6$ in decreasing order based on the pcR value of the agent profile; finally, it returns the first v product categories. The resulting list is the new $L2$, returned as the output of productOfInterest.

To preserve customer's privacy, we can note that TRES faces this issue in a simple and effective way. Specifically, it is the customer itself to decide which information has to be sent to a merchant agent. Obviously, the quality of the suggestions generated by TRES (and consequently of the Web presentation built on-the-fly by the merchant agent) is based on these suggestions and related to the precision degree to the customer's interests and preferences representation. Thus, less precise it is this representation and less attractive will be the suggestions (i.e., the Web presentation) proposed to the customer.

```
void Recommendations(customerAgent c, ListOfProductInstances L3,
                                      ListOfProductInstances L4) {
    ListOfProductCategories L1=extract_pc(m.UP.BD);
    send(L1,c.Ad);
    ListOfProductCategories L2=receive( );
    ListOfProductInstances L3=contentbased_pi(L2, m.UP.BD, y);
    ListsOfCustomersInterests PC[ ]=customersInterests(m.UP.BD, v);
    ListOfProductInstances L4=collaborativefiltering_pi(L2, PC[ ], m.UP.BD, z, x);
    return;
}

ListOfProductCategories productOfInterest(ListofProductCategories L1) {
    ListOfProductCategories L5=extract_pc(c.UP.BD);
    ListOfProductCategories L6=intersection_pc(L1, L5);
    ListOfProductCategories L2=select_pc(L6,v);
    return L2;
}
```

Fig. 6. The recommendation algorithm exploited in the generation of suggestions

3.2   The Computational Complexity of the TRES Recommendation Algorithm

Now we analyze how the time cost and the storage space required by TRES to perform the recommendation task are strictly related to the dimension of the multi-agent system, the dimension of the site catalogue and the dimension of the average number of product instances visited for each product category of the catalogue

In this analysis n denotes the dimension of the multi-agent system, represented by the number of the registered users, p is the dimension of the site catalogue, represented by the number of different concepts present in the catalogue, and q is the average number of product instances for each product category visited by the customer, it is computed by considering all the product categories present into the customer's profile.

More in detail, on the merchant side the computational cost to generate recommendation is due to the cost of performing the function `Recommendations`, described in Figure 6, that, in its turn, depends on the cost of the called functions `extract_pc`, `send`, `receive`, `contentbased_pi`, `customersInterests` and `collaborativefiltering_pi`. Preliminarily, with respect to the functions `send` and `receive` we consider the cost for an agent to send or to receive a message (denoted by C_T) as constant and independent from the particular agent. The first function called is `extract_pc` that computes the product categories of interest for the current customer by comparing the number of items in the site catalogue (p in the worst case) with those present in customer's profile (p in the worst case) obtaining a computational complexity of $\mathcal{O}(p^2)$. The task to compute content based suggestions is carried out by the function `contentbased_pi` that should examine p product instances for each one of the v product categories contained in $L2$. By considering that in the worst case $v \leq p$, the cost of this function is $\mathcal{O}(p \cdot q)$.

Then in `Recommendations` the function `customersInterests` is called that, as previously remarked, is the most expensive function of the TRES recommendation algorithm. This function considers in average q product instances for each product category selected by c from his/her profile. The number of product categories is equal to v in average (p in the worst case) for each registered user). As a result, the computational cost of this function in the worst case is $\mathcal{O}(n \cdot p \cdot q)$.

Finally, the function `collaborativefiltering_pi` selects for a customer c the most x interesting product instances (q in the worst case) for each one of the v product categories (p in the worst case) selected by c and considered also from the z most similar agents (n in the worst case). The agent similarity between the customer c and the other customers (n in the worst case) is computed by comparing the selected product categories (p in the worst case). Thus in the worst case the cost of `collaborativefiltering_pi` is $\mathcal{O}(n \cdot p \cdot q + n \cdot p^2)$.

By considering that the number of product instances associated with a seller is significantly greater than the number of product categories included in the catalogue, but that the product instances visited by a customer are a subset of the whole number of instances, it is reasonable to assume that the value of q is greater or equal to p. In this case, the computational cost of the function `Recommendations` results mainly dependent on the cost of the function `customersInterests` and it is $\mathcal{O}(n \cdot p \cdot q)$. The same considerations can be carried out also for the storage cost that is $\mathcal{O}(n \cdot p \cdot q)$, since it is mainly due to the need of storing all the customers' profiles that in the worst case depends on the number of registered users, the number of product categories in the catalogue and the number of product instances for each product category considered.

On the customer side the computational cost due to `productOfInterest` is $\mathcal{O}(p^2)$. In fact, the functions `extract_pc` and `intersection_pc` have a computational cost, that in the worst case is of $\mathcal{O}(p^2)$, while the cost of the function `select_pc` depends on the cost of the sort algorithm exploited therein that can be assumed as $\mathcal{O}(p \cdot logp)$. Moreover, on the customer side the storage cost depends only on the number of product categories, thus the global cost is $\mathcal{O}(p)$.

4   Related Work

Recommender systems are generally adopted in several EC sites (e.g., Amazon, CDNOW, GroupLens, MovieLens, etc. [6]) and a large number of models and architectures are based on software agents. In this section, we describe some past approaches that in the generation of the recommendations implement both content-based and collaborative filtering techniques [33–36] and appear as the closest to the proposal presented in this paper. At the end of the current section, differences and similarities with TRES will be pointed out. For a more complete account about this matter the interested reader might refer to the considerable number of surveys that have investigated the state of the art [3–7, 37, 38].

The multi-agent system IMPLICIT [39] uses a search engine (i.e., Google) while personal agents reciprocal interacting in order to generate suitable recommendations by exploiting the notion of *Implicit Culture*. The search engine results are thus complemented with the recommendations produced by the agents. In IMPLICIT each user is assisted by a personal agent during his/her search to find Web links considered relevant and for discovering agents to contact from which obtaining relevant links. The Search behaviour consists of the Google search behaviour and of the Platform search behaviour, which comprises both the Internal and the External search behaviour. In the Google search behaviour the agent processes query to the Google search engine in order to obtain some suggestions for any entered keyword. In the Internal search are generated links, based on the past user actions monitored by his/her agent while in the External search are proposed the most suitable agents to contact, also exploiting a yellow pages service provided by the Directory Facilitator of the platform. When all the agents are contacted and the search behaviour queries suggested by the new agents have been performed, the system shows all the discovered links to the user.

Another multi-agent recommender system is MASHA (Multi Agent System Handling Adaptivity) [40] that, as main feature, considers in its suggestions also the characteristics of the device currently exploited by a user in his/her Web activities. In MASHA each user is associated with a *server agent* that builds a global user profile collecting by the user's *client agents*, each one associated with a user's device, their profiles, storing information about the user's behaviour when he/she uses that device. The *adapter agents*, associated with each Web site, exploit global profiles to generate personalized Web site presentations containing suggestions derived from the profiles of the current user and those of other users that have visited the site in the past by exploiting the same device.

Handy Broker [41] adopts an evolutionary ontology-based approach for mobile trading activities. The system assumes users as rational individuals able to evaluate the product relevance uniquely by means of tangible attributes (e.g., the price), while intangible attributes, such as the brand, are not considered. The Handy Broker agent generates its recommendations based both *(i)* on the whole user's history stored in a user's profile in which his/her preferences are witnessed by how many times a product has been selected and *(ii)* an evolutionary mechanism that allows an agent to integrate in its profile those of other users for performing, in such a way, an implicit collaborative filtering recommender stage. In [42] agents support both recommendation and negotiation phases. In particular, the recommendation part is designed to assist a consumer for products rarely purchased. The system builds its knowledge of the products by interfacing with domain experts and uses the acquired knowledge to calculate optimal products that match customers' preferences (provided by means of a questioner). Optimality of the products is computed by using a multi-attribute decision making method based on consumer's needs and features of products.

Moreover, in order to share the experiences of other consumers, a dynamic programming approach is used to exploit social information derived from previous consumer recommendations.

CBCF [27] (Content-Boosted Collaborative Filtering) exploits *(i)* a content-based predictor to process user data (as text documents), rated by each user in six classes of relevance, and *(ii)* a collaborative filtering approach adopting a neighborhood-based algorithm for choosing a subset of users similar to him/her in order to obtain personalized recommendations by exploiting a weighted combination of their ratings.

A Graph can model in an easy way the usage information, as for example in X-Compass [28], SUGGEST [43] and in [44]. More specifically, in X-Compass a XML-based agent is associated with a user U in order to suggest the Web pages potentially interesting for him/her. To this aim the agent monitors the Web pages visited by U and automatically builds and manages a *user profile* modeled with a graph. In this graph each node is associated to a concept of interest for U, its *Attraction Degree* and a *Key Set* representing the semantics of the concept, while each arcs stands for both *is-a* relationships and associative rules. Furthermore, the agent updates two lists containing the visited Web pages, ordered with a temporal access criterion, and their whole visit history, respectively. Likewise, SUGGEST supports user Web navigation by dynamically generating links to unvisited pages so evaluated attractive for him. A complete graph is adopted to model historical user's navigational information (users' sessions are identified by means of cookies stored on the client side) and it is handled by means of an incremental graph partitioning algorithm. In the graph the set of vertices contains the identifiers of the pages hosted on the Web server and the set of edges contains the weight $W_{ij} = N_{ij}/max\{N_i, N_j\}$, where N_{ij} is the number of sessions containing both pages i and j, N_i and N_j are the number of sessions containing the only page i and j, respectively. The graph is then partitioned by using a clustering algorithm (a version of the incremental connected components algorithm) in order to find groups of strongly correlated pages. Subsequently, the cluster having the largest intersection with the page window correspondent to the current session contributes to form the suggestion list.

In [44] a two-layer graph model is exploited, where nodes, intra-layer and inter-layer links of two graphs are respectively used to represent customers and products (each one in a different layer), transactions and similarities. This representation of user-products relationships is exploited to implement content-based, collaborative filtering and hybrid recommenders by using "direct retrieval" (DR), "association mining" (AM) and "High-degree association retrieval" ($HDAR$) methodologies. In the DR methodology: the content-based part is similar to that used in information retrieval where documents similar to the input queries are retried; collaborative filtering is based on users' similarity evaluated by using information extracted from the graph topologies; the hybrid section simply combines the other two techniques. The AM methodology uses transaction histories and associative rules, that obviously change for content-based and for collaborative filtering recommendations, while hybrid approach combines them.

The last method ($HDAR$) deals with data sparsity by transitively exploring layer topologies to search information related to the neighborhoods of each users and to define associative rules for generating content-based and collaborative filtering recommendations. Joining such rules it is possible to obtain a hybrid recommender.

5   Experiments

In this section, we present some experiments devoted to show the effectiveness of TRES to generate useful suggestions for supporting users during their B2C activities. The experiments presented below have been realized by using a TRES prototype, developed in JADE [45, 46]), able to completely implement all its features throughout all the W-CBB activities.

For this experiment session we have *(i)* built a family of 18 XML EC TRES compliant Web sites (Figure 7) by using the NAICS coding as common vocabulary, represented by a unique XML Schema, with 760 product instances belonging to 9 NAICS categories and *(ii)* monitored a set of real users in their B2C activities within the TRES framework. The first 9 sites has been used to obtain an initial profile of the customers' interests and preferences without to exploit any recommendation support and to determine the value of the ρ parameters. Based on such profiles and ρ values, the recommendations have been generated by the merchant agents relatively to the other 9 sites.

We have compared TRES, identified in the following by $TRES - ON$, with a modified version of this system, identified in the following by $TRES - OFF$ where users' interests and preferences are taken into account simply by means of the only number of access to a product instance (i.e., product category) without to consider the trading context, computing piR and pcR, using the following formulation:

$$piR = \frac{piR}{log_{10}(10+\Delta)} + 1$$
$$pcR = \frac{pcR}{log_{10}(10+\Delta)} + 1$$

where Δ has the same meaning explained in Section 2.2.

Moreover, these tests have included the comparison with $CBCF$ and $X - COMPASS$, that are both two content-based and collaborative filtering recommender systems exploiting an user's profile built monitoring the user's behaviour. They are, at the best of our knowledge, two of the most performative recommender systems, as highlighted by the experimental results described in [27, 28] that compare these systems with other well-known recommender systems. Also the $CBCF$ and X-COMPASS agents have been implemented in JADE by following the descriptions of the data structures and recommendation algorithms proposed in [27, 28], respectively.

To evaluate the results of the experiments, we have inserted in a list, called A, the product instances suggested by the merchant agent and in a list, called B, the corresponding customer's choices. The associated pairs in the two lists

Table 2. Performances of different recommendation algorithms (global/content-based/collaborative filtering)

		$TRES-ON$	$TRES-OFF$	$CBCF$	$X-Compass$
	Pre	0.597/0.508/0.412	0.461/0.372/0.325	0.503/0.418/0.374	0.421/0.376/0.287
	Rec	0.564/0.482/0.392	0.434/0.336/0.289	0.467/0.392/0.341	0.407/0.354/0.244
	F	0.580/0.495/0.402	0.442/0.353/0.306	0.484/0.405/0.357	0.367/0.365/0.264

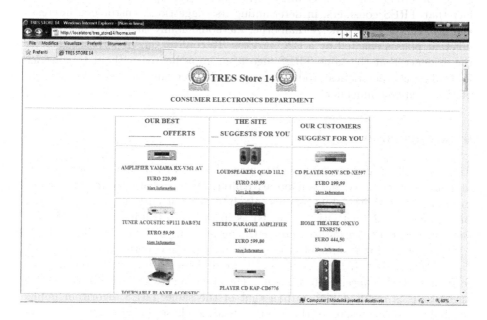

Fig. 7. The personalization of the home page of the site N.14

have been compared in order to measure the effectiveness of the generated suggestions. We have adopted standard performance metrics that are, precision, recall and F-measure [47] to compare the tested recommender approaches. More in detail, precision is defined as the share of the concepts actually visited by the user among those recommended by the system. Recall is the share of the concepts suggested by the system among those chosen by the user. F-Measure represents the harmonic mean between Precision and Recall. The three measures are definable as follows:

$$Pre(A(x)) = \frac{|A(x) \cap B(x)|}{|A(x)|} \quad , \quad Rec(A(x)) = \frac{|A(x) \cap B(x)|}{|B(x)|}$$

$$F(A(x)) = \frac{2 \times Rec(A(x)) \times Pre(A(x))}{Rec(A(x)) + Pre(A(x))}$$

The performance of the content-based and collaborative filtering components have been considered both in an integrated way and separately for all the tested recommenders. The parameters v, y, z and x of TRES have been set to 2, 3, 2 and 3, respectively (see Section 3). In terms of results (see Table 2) TRES has outperformed the other approaches chosen for the comparison. TRES shows with respect to CBCF, the best competitor as global, content-based and collaborative filtering performances, an improvement of precision of about 19, 22 and 10 percent and an improvement of recall of about a 21, a 23 and a 15 percent. Furthermore, from the results of Table 2 we can argue that the good performance of TRES, with respect to the other considered approaches, is due to the fact that TRES considers, in determining its suggestions, customers' interests and preferences more exactly than other methods mainly thanks to consider the different phases enacted by a B2C process. To confirm our hypothesis, there is the evidence that the same TRES algorithm implemented without to consider its evolute characteristics produces good results but only comparable with the other tested recommenders.

6   Conclusion

This paper illustrates a recommender system, called TRES, that is able to generate both content-based and collaborative filtering suggestions, to support traders in their B2C activities. Suggestions take into account customers' interests and preferences in B2C activities arranged as in the Web trading model and, as a consequence, contributing to increase the performances of the EC sites.

Some interesting results have been obtained by the experimental simulations carried out using a JADE-based prototypal implementation over an agent platform, appositely conceived to this aim. The effectiveness of TRES in generating suggestions is resulted higher than that of the same TRES algorithm implemented without considering the different trading phases enacted by a B2C process and also higher than that of other two competitor recommenders.

In the next future, further developments are expected from the introduction of different behavioural models acting in the B2C area that might contribute to build a more detailed customer's profile in order to generate more precise suggestions.

References

1. Kauffman, R.J., Walden, E.A.: Economics and Electronic Commerce: Survey and Directions for Research. International Journal of Electronic Commerce 5(4), 5–116 (2001)
2. Zwass, V.: Electronic Commerce and Organizational Innovation: Aspects and Opportunities. International Journal of Electronic Commerce 7(3), 7–37 (2003)
3. Sarwar, B., Karypis, G., Konstan, J., Riedl, J.: Analysis of Recommendation Algorithms for E-Commerce. In: Proceedings of the 2nd ACM Conference on Electronic Commerce (EC 2000), pp. 158–167. ACM, New York (2000)
4. Schafer, J.B., Konstan, J.A., Riedl, J.: E-Commerce Recommendation Applications. Data Mining Knowledge Discovery 5(1-2), 115–153 (2001)

5. Burke, R.D.: Hybrid Recommender Systems: Survey and Experiments. User Modeling and User-Adapted Interaction 12(4), 331–370 (2002)
6. Montaner, M., Lopez, B., de la Rosa, J.L.: A Taxonomy of Recommender Agents on the Internet. Journal of Web Semantics (JWS) 19(4), 285–330 (2004)
7. Wei, K., Huang, J., Fu, S.: A Survey of E-Commerce Recommender Systems. In: Proceedings of the 13th International Conference on Service Systems and Service Management, pp. 1–5. IEEE Computer Society, Washington, DC (2007)
8. Nicosia, F.: Consumer Decision Processes: Marketing and Advertising Implications. Prentice Hall, New York (1966)
9. Engel, J.F., Blackwell, R.D., Miniard, P.W.: Consumer Behaviour. International ed. The Dryden Press, London, UK (1995)
10. Nissen, M.E.: The Commerce Model for Electronic Redesign. J. of Internet Purchasing 1(2) (1997), http://www.arraydev.com/commerce/JIP/9702--01.htm
11. Feldman, S.: The Objects of the E-Commerce, Keynote speech at ACM 1999 Conference on OOPLSA, Denver (1999), http://www.ibm.com/iac/oopsla99-sifkeynote.pdf
12. Guttman, R.H., Moukas, A., Maes, P.: Agents as Mediators in Electronic Commerce. Electronic Markets 8(1) (1998)
13. He, M., Jennings, N.R., Leung, H.: On Agent-Mediated Electronic Commerce. IEEE Transaction Knowledge Data Engineering 15(4), 985–1003 (2003)
14. Palopoli, L., Rosaci, D., Ursino, D.: Agents' Roles in B2C e-Commerce. AI Communication 19(2), 95–126 (2006)
15. Maes, P.: Agents that Reduce Work and Information Overload. Communication of ACM 37(7), 30–40 (1994)
16. Hayes-Roth, B.: An Architecture for Adaptive Intelligent Systems. Artificial Intelligence 72(1-2), 329–365 (1995)
17. Wooldridge, M., Jennings, N.R.: Agent Theories, Architectures, and Languages: A Survey. In: Wooldridge, M.J., Jennings, N.R. (eds.) ECAI 1994 and ATAL 1994. LNCS, vol. 890, pp. 1–39. Springer, Heidelberg (1995)
18. Franklin, S., Graesser, A.C.: Is it an Agent, or Just a Program?: A Taxonomy for Autonomous Agents. In: Jennings, N.R., Wooldridge, M.J., Müller, J.P. (eds.) ECAI-WS 1996 and ATAL 1996. LNCS, vol. 1193, pp. 21–35. Springer, Heidelberg (1997)
19. Gilbert, D., Aparicio, M., Atkinson, B., Brady, S., Ciocarino, J., Grosof, B., O'Connor, P., Osisek, D., Pritko, S., Spagna, R., Wilson, L.: White Paper on Intelligent Agents. IBM Report, Zurich, Switzerland (1996)
20. Nwana, H.S.: Software Agents: An Overview. Knowoledge Engineering Review 11(3), 11–40 (1996)
21. Russell, S.J.: Rationality and Intelligence. Artificial Intelligence 94(1-2), 57–77 (1997)
22. Iglesias, C.A., Garijo, M., González, J.C.: A Survey of Agent-Oriented Methodologies. In: Papadimitriou, C., Singh, M.P., Müller, J.P. (eds.) ATAL 1998. LNCS (LNAI), vol. 1555, pp. 317–330. Springer, Heidelberg (1999)
23. Extensible Markup Language (XML) v.e. 1.1 (2010), http://www.w3.org/TR/2004/REC-xml11-20040204
24. Grosof, B.N., Labrou, Y.: An Approach to Using XML and a Rule-Based Content Language with an Agent Communication Language. In: Dignum, F., Greaves, M. (eds.) Agent Communication. LNCS (LNAI), vol. 1916, pp. 96–117. Springer, Heidelberg (2000)

25. O'Brien, P.D., Nicol, R.C.: FIPA Towards a Standard for Software Agents. BT Technology Journal 16(3), 51–59 (1998)
26. Foundation for Intelligent Physical Agents (FIPA) - ACL URL. FIPA ACL Message Structure Specif. (2010), http://www.fipa.org/specs/fipa00061/
27. Melville, P., Mooney, R.J., Nagarajan, R.: Content-boosted Collaborative Filtering for Improved Recommendations. In: Proceedings of the 18th National Conference on Artificial Intelligence, Edmonton, Canada, pp. 187–192. AAAI/IAAI (2002)
28. Garruzzo, S., Modafferi, S., Rosaci, D., Ursino, D.: X-Compass: An XML Agent for Supporting User Navigation on the Web. In: Andreasen, T., Motro, A., Christiansen, H., Larsen, H.L. (eds.) FQAS 2002. LNCS (LNAI), vol. 2522, pp. 197–211. Springer, Heidelberg (2002)
29. North America Industry Classifications (NAICS) (2010), http://www.census.gov/naics/2007/index.html
30. Extensible Markup Language (XML) Schema (2010), http://www.w3.org/XML/Schema
31. Asokan, N., Janson, P.A., Steiner, M., Waidner, M.: The State of the Art in Electronic Payment Systems. IEEE Computer 30(9), 28–35 (1997)
32. O'Mahony, D., Pierce, M., Tewari, H.: Electronic Payment Systems for E-Commerce, 2nd edn. Artech House, Norwood (2001)
33. Balabanovic, M., Shoham, Y.: Content-Based, Collaborative Recommendation. Communication of ACM 40(3), 66–72 (1997)
34. Aggarwal, C.C., Yu, P.S.: Data Mining Techniques for Personalization. IEEE Data Engineering Bulletin 23(1), 4–9 (2000)
35. Lawrence, R.D., Almasi, G.S., Kotlyar, V., Viveros, M.S., Duri, S.: Personalization of Supermarket Product Recommendations. Data Mining Knowledge Discovery 5(1/2), 11–32 (2001)
36. Tso, K.H.L., Schmidt-Thieme, L.: Evaluation of Attribute-Aware Recommender System Algorithms on Data with Varying Characteristics. In: Ng, W.-K., Kitsuregawa, M., Li, J., Chang, K. (eds.) PAKDD 2006. LNCS (LNAI), vol. 3918, pp. 831–840. Springer, Heidelberg (2006)
37. Wei, C.P., Shaw, M.J., Easley, R.F.: A Survey of Recommendation Systems in Electonic Commerce. In: E-Service: New directions in Theory and Practice. ME Sharpe, Armonk (2002)
38. Manouselis, N., Costopoulou, C.: Analysis and Classification of Multi-Criteria Recommender Systems. World Wide Web 10(4), 415–441 (2007)
39. Birukov, A., Blanzieri, E., Giorgini, P.: Implicit: A Recommender System that Uses Implicit Knowledge to Produce Suggestions. In: Workshop on Multi-Agent Information Retrieval and Recommender Systems at the 19th International Joint Conference on Artificial Intelligence (IJCAI 2005), Edinburgh, Scotland (2005)
40. Rosaci, D., Sarné, G.M.L.: MASHA: A Multi-Agent System Handling User and Device Adaptivity of Web Sites. User Modelling User-Adaptive Interaction 16(5), 435–462 (2006)
41. Guan, S., Ngoo, C.S., Zhu, F.: Handy broker: an Intelligent Product-Brokering Agent for m-Commerce Applications with User Preference Tracking. Electronic Commerce Research and Application 1(3-4), 314–330 (2002)
42. Lee, W.P.: Towards Agent-based Decision Making in the Electronic Marketplace: Interactive Recommendation and Automated Negotiation. Expert Systems with Applications 27(4), 665–679 (2004)

43. Silvestri, F., Baraglia, R., Palmerini, P., Serranó, M.: On-line Generation of Suggestions for Web Users. In: ITCC (1), pp. 392–397. IEEE Computer Society, Washington, DC (2004)
44. Huang, Z., Chung, W., Chen, H.: A graph model for e-commerce recommender systems. Journal of American Society Information Science Technology 55(3), 259–274 (2004)
45. Bellifemine, F., Poggi, A., Rimassa, G.: Developing Multi-Agent Systems with a FIPA-compliant Agent Framework. User Modelling User-Adaptive Interaction 12(4), 331–370 (2002)
46. Java Agent DEvelop. framew. (JADE) (2010), http://jade.tilab.com/
47. van Rijsbergen, C.J.: Information Retrieval. Butterworth (1979)

Simulation Analysis Using Multi-Agent Systems for Generalized Matching Pennies Games

Ichiro Nishizaki, Tsuyoshi Nakakura, and Tomohiro Hayashida

Department of System Cybernetics, Graduate school of Engineering, Hiroshima University,
1-4-1 Kagamiyama, Higashi-Hiroshima, 739-8527, JAPAN
{nisizaki,hayashida}@hiroshima-ua.c.jp

Abstract. In this chapter, to investigate the long-run behavior of players in the generalized matching pennies game, we employ an approach based on adaptive behavioral models. To do so, we develop an agent-based simulation system in which artificial adaptive agents have mechanisms of decision making and learning based on neural networks and genetic algorithms. We analyze the strategy choices of agents and the obtained payoffs in the simulations, and compare the predictions of Nash equilibria, the experimental data, and the results of the simulations with artificial adaptive agents. Moreover, we examine similarities between the behaviors of the human subjects in the experiments and those of the artificial adaptive agents in the simulations.

1 Introduction

There exists a unique Nash equilibrium composed of only mixed strategies in a generalized matching pennies game where the payoff structures of the row and the column players are asymmetric. However, it is reported that the behaviors of subjects in experiments are different from the prediction of the Nash equilibrium [26,19,20,13].

Ochs [26] conducts an experiment and its result is summarized as follows. Although for a symmetric matching pennies game the strategy choices of experimental subjects are coincident with the Nash prediction, for asymmetric games they are different from the Nash prediction. Observing that the game is repeated and therefore the experimental subjects revise their strategies by examining the outcomes of the games, Ochs classifies the behaviors of the subjects into several patterns from the viewpoint of learning of the subjects, and he claims that the models by McKelvey and Palfrey [19] and Roth and Erev [29] successfully explain the aggregated dynamics of the behaviors of the subjects in the experiment.

McKelvey and Palfrey [19] propose the quantal response equilibrium (QRE) model by introducing the error into the best response function from the idea that better responses are more likely to be observed than worse responses. As a special case of QRE model, they give the logit equilibrium model with a parameter of the error. They apply the QRE model to the three games used in the experiment by Ochs [26], and show that the behaviors of the experimental subjects can be well-accounted for by the QRE model.

McKelvey et al. [20] conduct an experiment in which the games used in the experiment by Ochs [26] are dealt with again, and they compare the Nash equilibrium, the

A. Håkansson & R. Hartung (Eds.): *Agent & Multi-Agent Syst. in Distrib. Syst.*, SCI 462, pp. 187–215.
DOI: 10.1007/978-3-642-35208-9_10   © Springer-Verlag Berlin Heidelberg 2013

random choice, the Nash model with noise which combines the Nash equilibrium with the random choice, and the QRE model. As a result of the comparison, they find that the QRE model accounts for the behavior of the subjects observed in experiments most accurately. They point out that the error rates of players are not always the same while it is assumed that the error in the QRE model is uniform. From this viewpoint, they extend the QRE model so as to be able to deal with the heterogeneity of the error, and show that the fitness of the experimental data is improved by the use of the extended model.

Roth and Erev [29] propose a simple model based on reinforcement learning in which the choice probabilities of strategies are revised on the basis of the obtained payoffs, they extend the model by introducing the two parameters of forgetting and similarity, and examine the effectiveness of the model by using the data observed in the experiments with a public good provision game, a market game, and an ultimatum bargaining game. Erev and Roth [11] verify the validity of the reinforcement learning model and its extended model through the experimental data of several games with unique equilibria including the games used in the experiment by Ochs [26].

Goeree et al. [14] extend the QRE model so as to express the risk attitude of a player through a utility function with a risk parameter, and estimate the values of the error and the risk parameters by using the data from the experiment on the generalized matching pennies games. They show that the proposed model with the error and the risk parameters can explain the experimental data very well.

Numerous experiments have been accumulated to examine human behaviors not only in generalized matching pennies games but also in the other games such as coordination games, ultimatum bargaining games, market entry games and so forth. Although in experimental studies, situations conforming mathematical models are formed in laboratories and experimental subjects are motivated by money, it can be pointed out that in such experiments with human subjects there exist limitations with respect to the number of trials, the number of subjects, variations of parameter settings and so forth.

In most mathematical models in economics and game theory, it is assumed that players are rational and maximize their payoffs, and they can discriminate between two payoffs with a minute difference. Such optimization approaches are not always appropriate for analyzing human behaviors and social phenomena, and models based on adaptive behavior can be alternatives to such optimization models. Recently as complements of conventional mathematical models, a large number of adaptive behavioral models have been proposed [3,4,7,10,11,18,28,29,31].

It is natural that behaviors of agents in simulation models is described by using adaptive behavioral rules, and simulations can be a promising approach to modeling situations where it is difficult to assume rational behaviors of decision makers in the strict sense. As concerns such approaches based on adaptive behavioral models, Holland and Miller [15] interpret most economic systems as complex adaptive systems, and point out that simulations using artificial societies with adaptive agents are effective for analysis of such economic systems. Axelrod [2] insists on the need for simulation analysis in social sciences, and states that purposes of the simulation analysis include prediction, performance, training, entertainment, education, proof and discovery.

In this chapter, we attempt to analyze the long-run behavior of players who repeatedly play the generalized matching pennies games by developing an agent-based simulation system for analyzing behaviors of agents in the games and conducting simulations with several settings. In our system for simulation of the generalized matching pennies games, the decision mechanism of an agent is based on a neural network with several inputs, and the agent chooses a strategy in accordance with the output of the neural network. The synaptic weights and thresholds characterizing the neural network are revised so as to obtain larger payoffs through a genetic algorithm, and then this learning mechanism develops agents with better performance.

If we put emphasis on only imitating the behaviors of experimental subjects, the learning mechanism based on natural selection is not necessarily required, and well-known technique such as the error back propagation algorithm with teacher signals would be more effective. However, to examine the behavior of players in the long run, we attempt to form a group of artificial adaptive agents acting autonomously such that agents obtaining larger payoffs hold dominant positions evolutionarily. Considering such an agent to be proxy for a human subject, we can estimate human behaviors in the generalized matching pennies game by conducting simulations extending over a long period of time and analyzing the results of the simulations. Moreover, we incorporate parameters of the risk attitude of agents and the error in decision making, and compare the predictions of Nash equilibria, the experimental data, and the results of the simulations with artificial adaptive agents learning by trial and error. We also examine similarities between the behaviors of the human subjects in the experiments and those of the artificial adaptive agents in the simulations. In section 2, we briefly review the generalized matching pennies games and the corresponding experiments with human subjects. Section 3 is devoted to explaining the agent-based simulation system with decision and learning mechanisms based on neural networks and genetic algorithms. In section 4, we examine the results of the simulations and compare the predictions of Nash equilibria, the experimental data, and the results of the simulations. Finally in section 5, we give a summary of the simulations and some concluding remarks.

2   Generalized Matching Pennies Games and Their Experiments

We briefly review the results of experimental studies with respect to generalized matching pennies games in this section. In the experiment by Ochs [26], the payoff table of a generalized matching pennies game shown in Table 1 is used. In this game, the row player has the two choices U and D and the column player also has the two choices L and R. When a strategy pair chosen by the row and the column players is (U,L) or (D,R), the row player receives a positive payoff of a or 1, respectively. When a strategy pair chosen by the players is (U,R) or (D,L), the column player receives a positive payoff of 1 in either case. For the payoff a of the row player with respect to the consequence (U,L), we assume that $a \geq 1$. When $a = 1$, the game is symmetric, and when $a > 1$, it is asymmetric. It is known that, in generalized matching pennies games, there does not exist any Nash equilibrium with pure strategies but there exists only a unique Nash equilibrium with strict mixed strategies.

Table 1. Generalized matching pennies game

		column player	
		L	R
row player	U	$(a,0)$	$(0,1)$
	D	$(0,1)$	$(1,0)$

Let p denote a probability that the row player chooses strategy U, and let q denote a probability that the column player chooses strategy L. Then, expected payoffs π_R and π_C of the row player and the column player, respectively, are represented by

$$\pi_R = apq + (1-p)(1-q), \tag{1}$$
$$\pi_C = (1-p)q + p(1-q), \tag{2}$$

and the corresponding Nash equilibrium is $(p^{\text{Nash}}, q^{\text{Nash}}) = (1/2, 1/(a+1))$. Because this game is not a zero-sum game, the maximin strategy is different from the Nash equilibrium. The maximin strategies of the row player and the column player are given by $\arg\max_p \min_q \pi_R(p,q)$ and $\arg\max_q \min_p \pi_C(p,q)$, respectively, and therefore the pair of the maximin strategies is $(p^S, q^S) = (1/(a+1), 1/2)$.

In the experiment by Ochs [26], three generalized matching pennies games with $a = 1, 4, 9$ are arranged, and a pair of experimental subjects repeat playing the game ten times. They can choose one of the two pure strategies or a mixed strategy. For a mixed strategy, a subject is asked the number of choosing U in ten games, and the computer system randomly chooses U or D in each game such that the total number of U is consistent with the number given by the subject. To control the effect of risk attitude of subjects, they are provided with not monetary payoffs but lottery tickets. The result of the experiment is summarized as follows.

(i) The frequency that the row player chooses U rises as the payoff a to the consequence (U, L) increases.
(ii) The frequency that the column player chooses L decreases as the payoff a increases.

The fact (i) of the result is not consistent with the prediction of Nash equilibrium, and the fact (ii) is not consistent with the prediction of the maximin strategy.

The relative frequencies in the experiments by Ochs [26], McKelvey and Palfrey [19] and McKelvey et al. [20] are shown in Table 2 as well as the Nash equilibrium prediction. In particular, the values in the column of Ochs [26] are estimates of the long-run

Table 2. Results of the experiments: the strategy choice probability (p,q)

	Ochs [26]	McKelvey and Palfrey [19]	McKelvey et al. [20]	Nash
$a = 1$	$(0.5015, 0.4819)$	——	——	$(0.5, 0.5)$
$a = 4$	$(0.5336, ——)$	$(0.542, 0.336)$	$(0.550, 0.328)$	$(0.5, 0.2)$
$a = 9$	$(0.6309, 0.3497)$	$(0.595, 0.258)$	$(0.643, 0.241)$	$(0.5, 0.1)$

steady-state relative frequency, and the estimate of the probability q for the game with $a = 4$ is not given.

Ochs [26] states that although the aggregate dynamics observed in the experiment cannot be accounted for neither by the prediction of Nash equilibrium nor by the security motivations, they are well-accounted for both by the QRE model by McKelvey and Palfrey [19] and by a learning model by Roth and Erev [29].

McKelvey and Palfrey [19] propose the QRE model by introducing the error into the best response function from the viewpoint that better responses are more likely to be observed than worse responses, and give the logit equilibrium model with a parameter λ of the error as a special case of QRE model. Let π_{ij} denote an expected payoff of player i who chooses the jth pure strategy. Then, the probability that player i chooses the jth pure strategy is given by

$$p_{ij} = \frac{e^{\lambda \pi_{ij}}}{\sum_k e^{\lambda \pi_{ik}}}. \tag{3}$$

If $\lambda = 0$, the player randomly chooses a pure strategy, and if $\lambda = \infty$, the player maximizes the expected payoff, i.e., the player chooses a strategy of the Nash equilibrium. The value of λ is estimated by using experimental data.

Roth and Erev [29] and Erev and Roth [11] propose a model based on reinforcement learning and try to account for the behavior observed in experiments. In their model, from experience of repeatedly playing the games, the probability of choosing a pure strategy is updated. Namely, the propensity $q_{ij}(t)$ for player i to choose the jth pure strategy at period t is updated both by the propensity $q_{ij}(t-1)$ at the previous period and by a reinforcement $R(x)$ of receiving a payoff x as follows:

$$q_{nj}(t) = \begin{cases} q_{nj}(t-1) + R(x) & \text{if } j = k \\ q_{nj}(t-1) & \text{otherwise.} \end{cases} \tag{4}$$

Then, the probability that player i chooses the jth pure strategy is given by

$$p_{ij}(t) = \frac{q_{ij}(t)}{\sum_k q_{ik}(t)}. \tag{5}$$

The model is extended by introducing two parameters of "forgetting" which means that the recent experience may play a larger role than the past experience and "similarity" which means that if a ceratin strategy is reinforced, similar strategies are also reinforced.

Goeree et al. [14] extend the logit equilibrium model by incorporating risk attitude of players and substituting the expected utility u_{ij} for the expected payoff π_{ij} in (3). Let x be a payoff, and for example, it is assumed that the utility is represented by $u(x) = x^{1-r}$. They estimate the parameters λ and r of the error and the risk attitude by using the maximum likelihood estimation and show that their model well accounts for the behavior of subjects observed in experiments.

3   Agent-Based Simulation System

3.1   Related Works and Implementation of Agents

As we mentioned in the previous section, to account for the experimental data, the QRE model and the reinforcement learning model are developed, and the rationality of players' behavior is relaxed in those models. In particular, an alternative to the rational behavior of players is thought to be the adaptive behavior in the reinforcement learning model. In this study, we also suppose that human behavior is adaptive, and employ a simulation model which is a natural framework to implement the adaptive behavior of players.

Before describing our simulation system, we briefly review related studies on multi-agent systems for analyzing social behaviors of people and social phenomena. For the iterated prisoner's dilemma game, Axelrod [1] examines the effectiveness of strategies generated in an artificial social system, in which agents endowed with strategies are adaptively evolved by using a genetic algorithm. Dorsey et al. [7] employ an artificial decision making mechanism using neural networks to imitate the decision making of auctioneers, and compare artificial agents' behavior with that of real auctioneers which often deviate from Nash equilibria. To estimate bid functions of bidders, i.e., to establish the appropriate weights of a neural network, they employ the genetic adaptive neural network algorithm based on a genetic algorithm instead of the error back propagation algorithm which is the most commonly used method. Andreoni and Miller [3] use genetic algorithms to model decision making in auctions. In a way similar to the approach of Dorsey et al. [7], they compare decisions of artificial adaptive agents with the decisions observed in the experiment with human subjects, and find that they resemble each other. Erev and Rapoport [10] investigate a market entry game by using an adaptive learning model based on reinforcement learning proposed by Roth and Erev [29]. Rapoport et al. [28] also deal with market entry games. They compare the decisions observed in experiment with human subjects with those of artificial adaptive agents with a learning mechanism using reinforcement learning, and analyze behavioral patterns on the aggregate level.

Leshno et al. [18] consider equilibrium problems in market entry games through agent-based simulations with a decision making mechanism of agents based on neural networks, and the neural networks are trained not by some teacher signals but by the outcomes of games. They compare the results of the simulations with the results of the experiments with human subjects conducted by Sundali et al. [31], and find some similarities between phenomena of the simulations and the experiments. Nishizaki et al. [24] investigate the effectiveness of a socio-economic system for preserving the global commons by simulation analysis. Nishizaki [22] propose a general framework of decision making and learning mechanism based on neural networks and genetic algorithms for adaptive artificial agents in a multi-agent system, and Nishizaki et al. [23] apply this mechanism to multi-agent system for analyzing development of social norms. A number of attempts have been made for performing multi-agent based simulations and developing the related techniques underlying the ideas from distributed artificial intelligence and multi-agent systems [9,6,5,8,21,4,25,27,30].

As we mentioned above, reinforcement learning, learning classifier systems, neural networks, genetic algorithms, etc. are applicable as mechanisms for decision making and learning of adaptive artificial agents. To select the most appropriate mechanism, let us consider factors which have some influence on the behavior of human subjects in the experiments for generalized matching pennies games. It is supposed that the behavior of experimental subjects is primarily affected by the payoffs arising from their own actions and especially the payoff received at the latest period has the strongest impact. As for the effect of past experiences, it is natural to consider that the impact of the past payoffs varies among subjects. As Duffy and Feltovich [12] suggest, the behaviors or payoffs of the others are thought to affect the actions of self. Moreover, as McKelvey and Palfrey [19] incorporate the error into the best response function from the viewpoint that better responses are more likely to be observed than worse responses, the error in decision making should be taking into account. From the result of Goeree et al. [14], the risk attitude is an important issue for analyzing the behaviors of experimental subjects.

We give characteristics of three mechanisms for decision making and learning of adaptive artificial agents. In reinforcement learning, learning of an agent is effected by revising the selection probabilities of strategies according to the outcomes of actions selected by the agent. The learning classifier systems which consist of multiple "if-then" rules receive information from the environment and classify states of the environment; and the action given by the if-then rule matching a state of the environment is performed. According to rewards from the environment, a set of the rules is evolved. In the learning system with decision making and learning mechanisms based on neural networks and genetic algorithms developed by Nishizaki [22], an action of an agent is determined by an output of the neural network which is a nonlinear function of multiple inputs, and the parameters of the neural network are revised according to the outcomes of actions of the agent in the framework of genetic algorithms.

We now evaluate the applicabilities of these mechanisms for artificial agents in a multi-agent system for analyzing the behaviors of players in generalized matching pennies games. For reinforcement learning, it is difficult to relate actions or payoffs of the opponents to decisions of self and to incorporate all the aforementioned factors affecting players' behaviors. In generalized matching pennies games played repeatedly, it is thought that behaviors of players vary gradually. From this viewpoint, learning classifier systems seem to be not appropriate because actions of agents by them are determined discrete outputs based on if-then rules and therefore it is difficult to derive successive changes of an agent's actions. In contrast, the decision making and learning mechanism based on neural networks and genetic algorithms outputs continuous real numbers, and then the behaviors of agents vary in a continuous manner responding to the slow changes of the circumstance. Moreover, as shown in the next subsection, we can easily utilize a wide variety of information as input data of neural networks into this mechanism, and by revising the output data of the neural networks, the risk attitude and the errors can be incorporated. From these reasons, we employ this mechanism, and decisions of an artificial agent are determined by the outputs of the neural network and it can learn so as to obtain larger payoffs through a genetic algorithm.

3.2   Decision Making by a Neural Network

An agent corresponds to a neural network which is characterized by synaptic weights between two nodes in the neural network and thresholds which are parameters for the output function of nodes. Because a structure of neural networks is determined by the number of layers and the number of nodes in each layer, an agent is prescribed by these numbers of the parameters if they are fixed. In our model, we form a string consisting of these parameters which identify an agent, and think of the string as a chromosome in an artificial genetic system. Each agent in the population repeatedly plays a generalized matching pennies game, and the population evolves into that of agents with larger payoffs. A structure of neural networks and a chromosome are depicted in Fig. 1.

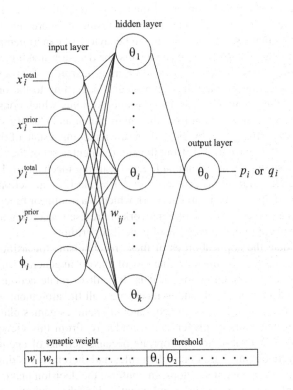

Fig. 1. Structures of a neural network and a chromosome

An output of a neural network is determined by a vector of inputs, the synaptic weights, and the thresholds. The inputs of the neural network include not only the payoffs of self obtained in the past games but also the payoffs of an opponent. To be more specific, the input vector consists of the following five values.

(i)   The total payoff obtained by agent i for all the past games: x_i^{total}
(ii) The payoff obtained by agent i at the prior game: x_i^{prior}
(iii) The total payoff obtained by the opponent for all the past games: y_i^{total}

(iv) The payoff obtained by the opponent at the prior game: y_i^{prior}

(v) The rate of carrying over a payoff to the next period: ϕ_i

The output of a neural network is interpreted as a probability that the agent chooses strategy U if the agent is the row player or strategy L if the agent is the column player.

Let $x_i(j)$ be a payoff of player i at period j. Then, the total payoff of player i at period t is represented by

$$x_i^{\text{total}}(t) = \sum_{j=1}^{t} \phi_i^{t-j} x_i(j). \tag{6}$$

On the assumption that a decision of the agent is affected by the payoff obtained before, though the previous payoff is reduced by $(1 - \phi_i)$, the payoff two periods ago is reduced by $(1 - \phi_i)^2$, and so on, the total payoff (6) is used as the first input value of the neural network. In the extended reinforcement model by Erev and Roth [11], a similar parameter is incorporated. In the experiment by Ochs [26], it is found that there exist some experimental subjects who take an outcome of the previous game into account to make a decision at the current period. From this viewpoint, we employ the payoff obtained at the prior game as the second input value of the neural network. Furthermore, Duffy and Feltovich [12] suggest the influence of the behavior or payoffs of the others on the strategy choices of self, and in our model the payoffs of an opponent as well as the payoffs of self are incorporated as the third and the fourth input values of the neural network. As above stated, the value of $(1 - \phi_i)$ can be interpreted as the rate of forgetting. Then, in our model, the fifth input value ϕ_i to the neural network is fixed for each agent and it is randomly assigned to each agent at the beginning of the simulation. Through the process of the genetic algorithm, it follows that the values of ϕ_i possessed by agents who are evolutionarily dominant survive in the long term.

Let z_l^k, $l = 1, \ldots, m^k$ denote inputs to node k, and let w_l^k, $l = 1, \ldots, m^k$ denote the corresponding synaptic weights. Let θ^k be the threshold of node k, and then an output o^k of node k is represented by

$$o^k = f\left(\sum_{l=1}^{m^k} w_l^k z_l^k - \theta^k \right), \tag{7}$$

where an output function f is monotonously increasing, and in our model we employ the sigmoid function $f(z) = 1/(1 + \exp(-z))$.

As we mentioned above, there are the five units in the input layer and there is one unit in the output layer. Let m be the number of units in the hidden layer. Then because the number of synaptic weights becomes $6m$, the neural network corresponding to an agent can be determined by the synaptic weights w_l, $l = 1, \ldots, 6m$ and the thresholds θ_l, $l = 1, \ldots, m+1$.

In general, for a given data set of vectors of inputs and outputs, the synaptic weights and the thresholds are adjusted by using a learning method such as the error back propagation, and a nonlinear function which represents the relation between the inputs and the outputs is identified. In the context of this study, by adjusting the synaptic weights and the thresholds by using data from the experiments, the neural network may imitate the behavior of human subjects. However, from the viewpoint of the agents' adaptive behavior based on trial and error, we do not take such an approach, but in contrast the synaptic weights and the thresholds are adjusted through the genetic algorithm in which the initial population evolves into the population of agents obtaining larger payoffs.

3.3   Procedure of the Game and Learning by a Genetic Algorithm

Agents repeatedly play the game, and accumulate the payoffs obtained in the iterated games. In our artificial genetic system, because the fitness of an individual, which means an agent, is based on its payoffs, agents obtaining larger payoffs are likely to survive.

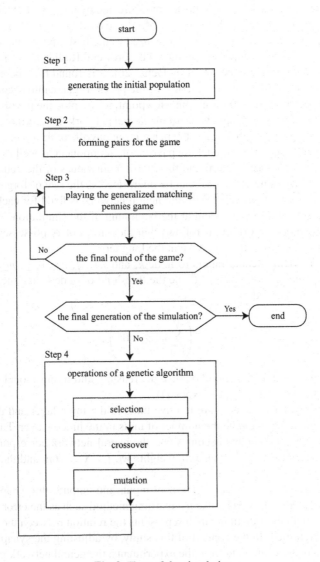

Fig. 2. Flow of the simulation

Because it is thought that the row and the column players differ in distributions of parameter values of the neural network after the learning, in our simulation model, we separately arrange two groups of agents who play the role of the row and the column players in the generalized matching pennies games. One agent is selected from each group, and the two agents make a pair for playing the game. Each pair repeat playing the game for a given number of times. For example, let the number of iteration be 100. If the output of the neural network is 0.7, the row agent chooses U as often as 70 times and D 30 times in all. This procedure is similar to that of the experiment by Ochs [26] in the case where a subject chooses a mixed strategy.

The structure of the simulation model is shown in Fig. 2, and the procedure of the simulation is summarized as follows.

Step 1. Generating the initial population: Let the number of the row or the column agents be n. Then, the initial population of $2n$ agents is formed by assigning random numbers to the parameters of the synaptic weights and the thresholds characterizing the neural network.

Step 2. Forming pairs for the game: One agent from the group of the row agents and one from the group of the column agents are randomly chosen, and then one pair for playing the game is formed. This procedure is repeated n times, and eventually, n pairs are made in all.

Step 3. Playing the game: Each agent chooses a strategy in accordance with the output of the neural network, and each pair repeatedly play the generalized matching pennies game for a given number of rounds. For each agent, the payoff x_i, which is the sum of payoffs in all the rounds, is recorded at the final round. If the generation number of the genetic algorithm reaches the maximum, the simulation is terminated.

Step 4. Performing genetic operations: The following genetic operations are performed to each of the chromosomes corresponding to the agents. After the genetic operations are performed and the population of the next generation is formed, return to Step 3.

    Step 4-1. Reproduction: The fitness f_i of each agent is the same as the payoff x_i obtained in the present generation[1]. As a reproduction operator, the roulette wheel selection is adopted. By a roulette wheel with slots sized by the probability $p_i^{selection} = f_i / \sum_{i=1}^{2n} f_i$, each chromosome is selected into the next generation.

    Step 4-2. Crossover: A single-point crossover operator is applied to any pair of chromosomes with the probability of crossover p_c. Namely, a point of crossover on the chromosomes is randomly selected and then two new chromosomes are created by swapping subchromosomes which are the right-hand side parts of the selected point of crossover on the original chromosomes. A new population is formed by exchanging the population in which the crossover operation is executed for the present generation with a given probability G. The probability G is called the generation gap. An agent keeps the history of obtaining payoffs in the past games, and the payoffs are divided between two offsprings in the proportion of sizes of the swapped subchromosomes.

---

[1] When we take into account the risk attitude of agents, the utility (8) defined in the next subsection is used for the fitness f_i instead of the payoff x_i.

Step 4-2. Mutation: With a given small probability of mutation p_m, each gene which represents a synaptic weight or a threshold in a chromosome is randomly changed. The selected gene is replaced by a random number in $[-1, 1]$.

3.4 Risk Attitude and Error

As McKelvey and Palfrey [19] propose the QRE model by introducing the error into the best response function, it is natural to suppose that there exist some errors in decisions by players. Moreover, Goeree et al. [14] incorporate the risk attitude of players into the QRE model and show that their model explains the experimental data very well. Sharing these viewpoints, we introduce the risk attitude and the error into the decision making and learning mechanism of our simulation model, and by analyzing the data from the simulations, we examine the influence of them on the behavior of agents.

3.4.1 Risk Attitude of Players

Goeree et al. [14] incorporate the risk attitude of players into the QRE model by McKelvey and Palfrey [19] and show that their model explains the experimental data successfully. Let x denote a payoff. A utility of an agent is given by x^{1-r}, and the utility of the agent is used as the fitness in the genetic algorithm. If $r > 0$, the risk attitude of the agent is risk averse; if $r < 0$, it is risk prone; and if $r = 0$, it is risk neutral. We can examine the influence of risk on choosing strategies by analyzing data from the simulations with several values of r distributed from risk averse to risk prone. Because the maximal payoff of the row agents is not the same for the values of a, the utility of the row agent is normalized, and the utility functions of the row and the column agents, respectively, are represented by

$$u_{\text{row}}(x; a, r) = \frac{x^{1-r}}{a^{1-r}}, \text{ and } u_{\text{col}}(x; r) = x^{1-r}. \tag{8}$$

In Fig. 3, the utility functions are shown for $r = -0.5, 0.0, 0.5, 0.8$. The left-hand side graph is the utility functions of the column player, and the right-hand side graph is the utility function of the row player in the case of $a = 9$.

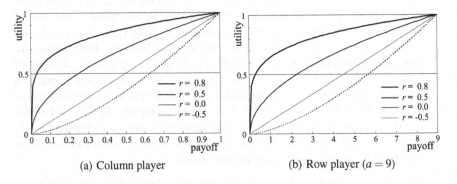

(a) Column player   (b) Row player ($a = 9$)

Fig. 3. Utility functions

As seen in Fig. 3, to obtain the same utility, a risk prone player must gain a larger payoff than that of a risk averse player. Moreover, a marginal utility of a risk averse player is larger than that of a risk prone player when the payoff is small, but this is reversed when the payoff is large. By using a value of the utility function as the fitness in the genetic algorithm, we can examine the influence of the risk attitude of agents on their behavior in our simulation model.

3.4.2   Error in Decision Making

The error in decision making of agents can be implemented by manipulating the output of the neural network, and we employ two methods: (i) simply adding an error term to the output, and (ii) modifying the output so as to approach the random choice.

The first method adding the error term is characterized by a magnitude e_1 and a direction α of the error. Let p_i denote the probability that agent i chooses U, and let q_j denote the probability that agent j chooses L. Then, the revised probabilities p'_i and q'_j including the error are represented by

$$p'_i = p_i + e_1 \cos \alpha, \tag{9}$$
$$q'_j = q_j + e_1 \sin \alpha, \tag{10}$$

where $0 \le \alpha < 2\pi$, and a value of the direction α is randomly determined for each game. If $p'_i > 1$ and $q'_j > 1$, they are modified by setting $p'_i = 1$ and $q'_j = 1$, respectively, and if $p'_i < 0$, $q'_j < 0$, they are modified by setting $p'_i = 0$ and $q'_j = 0$, respectively. As the value of e_1 becomes larger, the modified values p'_i and q'_j are away from the original values p_i and q_j and the error grows larger.

From a viewpoint that any strategy is selected with a similar probability if the error is large, in the second method the strategy choice probability approaches the random choice as the degree of error increases. This method is characterized by a magnitude e_2 of the error, and the original probabilities are modified as follows:

$$p'_i = \frac{1}{2}\{e_2 \log\{(1 - g(p_i))/g(p_i)\} + 1\}, \tag{11}$$
$$q'_j = \frac{1}{2}\{e_2 \log\{(1 - g(q_i))/g(q_i)\} + 1\}, \tag{12}$$

where $g(x) = x(g_1 - g_2) + g_2$, $g_1 = 1/(1 + \exp(1/e_2))$, and $g_2 = 1/(1 + \exp(-1/e_2))$. For $e_2 = -0.5, -0.2, -0.1$, the modified probabilities are depicted in Fig. 4.

In Fig. 4, the horizontal and the vertical axes mean the original and the modified probabilities, respectively, and it is found that the modified probabilities are close to 0.5, i.e., the random choice as the value of e_2 becomes larger. Because low probabilities are modified to be overweighed and high probabilities are modified to be underweighed, this modification can be also interpreted as a function of probability weighting [17,16].

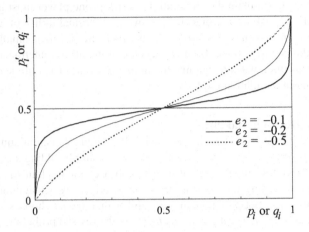

Fig. 4. Modified probabilities in the second method

4   Results of Simulations

4.1   Description of Simulations

After giving the results of simulations of symmetric and asymmetric games, we examine the influence of asymmetry in the payoff structure of the games on the behavior of artificial agents by varying the value of a which is a payoff of the row player. Furthermore, we investigate the effects of the risk attitude of players and the error in decision making by varying the values of r and e (e_1 and e_2), which are parameters of the risk attitude and the error, respectively. Then, we arrange the following five simulations.

 (i) Simulation $a = 1$: for examining the property of symmetric games ($a = 1$)
 (ii) Simulation $a = 9$: for examining the property of asymmetric games ($a = 9$)
(iii) Simulation Asymmetry: for examining the influence of asymmetry in the payoff structure of the games
(iv) Simulation Risk: for examining the influence of the risk attitude of players
 (v) Simulation Error: for examining the influence of the error in decision making

The following conditions are common in all the simulations.

- The number of units in the hidden layer of a neural network is ten.
- A sigmoid function $f(x) = 1/(1 + \exp(-x))$ is employed as an output function of a unit in a neural network.
- The numbers of agents for both of the row and the column players are 100 in the population.
- For parameters of the genetic algorithm, the probabilities of crossover and mutation are $p_c = 0.5$ and $p_m = 0.01$, respectively, and the generation gap is $G = 0.5$.
- The maximum number of generations of one run is 1000.
- Each agent plays the game with the fixed opponent 100 times in each generation.

- The execution of a simulation with the same condition is repeated 10 times, and the results of simulations are shown by the averages of 10 runs.
- At the beginning of a run, genes, i.e., synaptic weights or thresholds, in a chromosome are set at random numbers in the interval $[-1,1]$. In the mutation of the genetic algorithm, a selected gene is also replaced by a random number in $[-1,1]$.

4.2 Simulation $a = 1$ (Symmetric Case)

First, we focus on a game where the payoff structure of the row and the column players is symmetric, i.e., a matching pennies game with $a = 1$, and examine the strategy choice probabilities and the payoffs obtained by agents.

In this simulation, agents are assumed to be risk neutral and to make a decision with no error, and then the parameters of risk attitude and the error are set at $r = 0$ and $e_1 = 0$, respectively. The probability p that the row player chooses strategy U and the probability q that the column player chooses strategy L in each generation are shown in Fig. 5 together with the probabilities in Nash equilibrium. These probabilities are the averages of all agents. Similarly, the payoffs of the row and the column players are shown in Fig. 6 together with the payoffs in Nash equilibrium. The result of Simulation $a = 1$ is summarized in Table 3, and the averages of all the agents are shown as well as the average of the top ten pairs, which are ten pairs of agents with largest sum of the utilities of the row and the column agents $u_{\text{row}} + u_{\text{col}}$. In Simulation $a = 1$, from the definition of the utility function (8), $u_{\text{row}} = x_{\text{row}}$ and $u_{\text{col}} = x_{\text{col}}$, where x_{row} and x_{col} denote the payoffs of the row player and the column player, respectively.

In the earlier generations of the simulation, on average agents randomly choose their strategies because synaptic weights and thresholds of their neural networks are randomly determined. Therefore, as seen in Figs. 5, 6 and Table 3, the initial probabilities of p and q that the row and the column agents choose strategies U and L, respectively, are $(p,q) = (0.49,0.50)$, and the corresponding average payoffs of the row and the column agents are $(x_{\text{row}},x_{\text{col}}) = (0.5,0.5)$. It is observed that Nash equilibrium of this game is $(p^{\text{Nash}},q^{\text{Nash}}) = (0.50,0.50)$, and the result of Simulation $a = 1$ is very close to the prediction of Nash equilibrium. Although there exists an oscillation in the transition of p or q, the probabilities of strategy choices are $(p,q) = (0.51,0.50)$ and the pair of

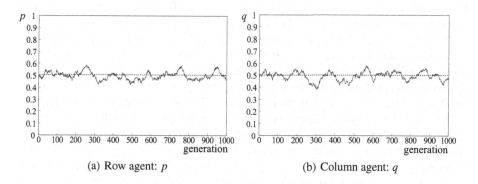

(a) Row agent: p (b) Column agent: q

Fig. 5. Probabilities p and q: $a = 1$

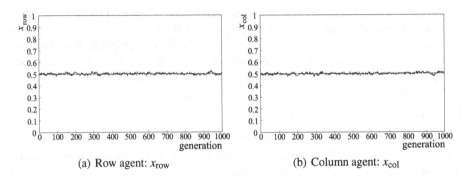

(a) Row agent: x_{row} (b) Column agent: x_{col}

Fig. 6. Payoffs x_{row} and x_{col}: $a = 1$

Table 3. Result of Simulation $a = 1$

generation	1–10	45–54	95–104	295–304	495–504	695–704	701–1000	Nash
p	0.49	0.53	0.53	0.50	0.44	0.51	0.51	0.50
q	0.50	0.50	0.51	0.40	0.50	0.52	0.50	0.50
x_{row}	0.50	0.50	0.50	0.51	0.50	0.50	0.50	0.50
x_{col}	0.50	0.50	0.51	0.49	0.50	0.50	0.50	0.50
p^{top}	0.48	0.52	0.54	0.50	0.44	0.50	0.50	—
q^{top}	0.50	0.50	0.50	0.39	0.50	0.53	0.50	—
$x_{\text{row}}^{\text{top}}$	0.50	0.50	0.50	0.51	0.50	0.50	0.50	—
$x_{\text{col}}^{\text{top}}$	0.51	0.50	0.50	0.49	0.50	0.50	0.50	—

average payoffs are $(x_{\text{row}}, x_{\text{col}}) = (0.5, 0.5)$ in the final 300 generations of the simulation. Thus, it follows that the result of Simulation $a = 1$ is almost coincident with the prediction of Nash equilibrium as well as the experiments with human subjects.

4.3 Simulation $a = 9$ (Asymmetric Case)

In this subsection, we deal with a game where the payoff structure of the row and the column agents is asymmetric and the parameter a of a payoff for the row player is set at $a = 9$, and examine the strategy choice probabilities and the payoffs obtained by agents.

In this simulation, agents are assumed to be risk neutral and to make a decision with no error, and then the parameters of risk attitude and error are set at $r = 0$ and $e_1 = 0$, respectively. The probability p of the row agent and the probability q of the column agent are shown in Fig. 7 together with the probabilities in Nash equilibrium, and the payoffs of the row and the column players are shown in Fig. 8 together with the payoffs corresponding to the Nash prediction and the results of the experiments with human subjects, which are depicted by a dashed line and a gray rectangle, respectively. The result of the experiments with human subjects is represented by the interval of the minimum and the maximum in Table 2 showing the results by Ochs [26], McKelvey and Palfrey [19] and McKelvey et al. [20]. In Fig. 9, the trajectories of the probability pair of strategy choices (p,q) and the payoff pair (x_{row}, x_{col}) are shown, and for comparison, the Nash prediction and the results of the experiments with human subjects are also depicted. The trajectory of (p,q) starts from $(0.5, 0.5)$ in the center of the figure and moves to the upper left of the figure, and the trajectory of (x_{row}, x_{col}) starts from $(2.5, 0.5)$ in the center of the figure and rotates to the upper left of the figure in a counterclockwise direction. In Fig. 8 and (b) of Fig. 9, the averages of the top ten pairs of agents are also provided. Moreover, the result of Simulation $a = 9$ is summarized in Table 4.

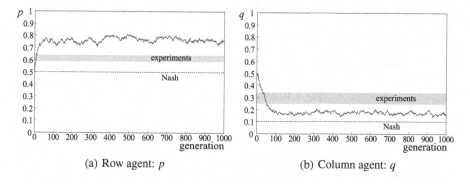

(a) Row agent: p (b) Column agent: q

Fig. 7. Probabilities p and q: $a = 9$

Because synaptic weights and thresholds of neural networks are randomly determined at the beginning of the simulation, as seen in Fig. 7 and (a) of Fig. 9, the probability pair of strategy choices (p,q) begins from $(p,q) = (0.50, 0.50)$ which is regarded as the random choice, and then the row and the column agents receive the corresponding payoffs $(x_{row}, x_{col}) = (2.50, 0.50)$.

At the initial point $(p,q) = (0.50, 0.50)$, although the payoffs of the column agents do not change if the column agents unilaterally increase or decrease the probability q, the row agents can obtain larger payoffs by increasing the probability p. From this reason, we can understand the observation that the row agents begin to choose strategy U more often and consequently to meet the action of the row agents, the column agents begin to decrease the frequency of choosing strategy L, i.e., the probability q. Namely, as seen Fig. 7, the row agents increase p until around generation 50, and the column agents decrease q until around generation 100. Thereafter, the probabilities p and q oscillate between the intervals $p \in [0.70, 0.81]$ and $q \in [0.14, 0.20]$, respectively. The averages of the probabilities after generation 100 are $(\bar{p}, \bar{q}) = (0.766, 0.173)$, and the corresponding payoffs are $(\bar{x}_{row}, \bar{x}_{col}) = (1.395, 0.673)$. For the top ten pairs, the probabilities p and q oscillate between the intervals $p \in [0.71, 0.87]$ and $q \in [0.13, 0.24]$, respectively, and the averages of the probabilities after generation 100 are $(\bar{p}^{top}, \bar{q}^{top}) = (0.785, 0.182)$. The top ten pairs of agents are more quick and excessive in action than the average behavior of all the agents. The corresponding payoffs are $(\bar{x}_{row}^{top}, \bar{x}_{col}^{top}) = (2.170, 0.742)$, which are close to the Pareto frontier, as seen in (b) of Fig. 9.

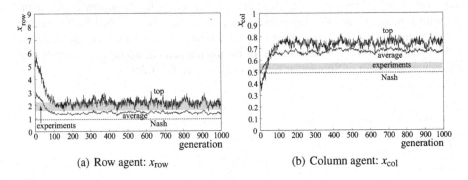

(a) Row agent: x_{row} (b) Column agent: x_{col}

Fig. 8. Payoffs x_{row} and x_{col}: $a = 9$

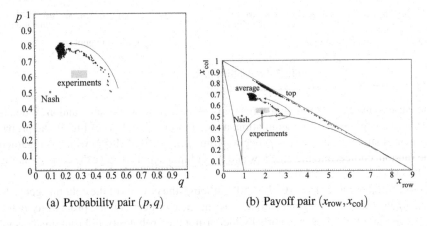

(a) Probability pair (p, q) (b) Payoff pair (x_{row}, x_{col})

Fig. 9. Trajectories of the probability pair (p, q) and the payoff pair (x_{row}, x_{col}): $a = 9$

Table 4. Result of Simulation $a = 9$

generation	1–10	45–54	95–104	295–304	495–504	695–704	701–1000	Nash
p	0.57	0.76	0.75	0.77	0.80	0.78	0.77	0.50
q	0.48	0.27	0.18	0.18	0.18	0.17	0.17	0.10
x_{row}	2.71	1.99	1.44	1.44	1.45	1.37	1.37	0.90
x_{col}	0.50	0.62	0.65	0.67	0.69	0.69	0.68	0.50
p^{top}	0.59	0.78	0.77	0.79	0.82	0.80	0.78	—
q^{top}	0.49	0.27	0.20	0.20	0.18	0.18	0.18	—
x_{row}^{top}	5.36	3.47	2.32	2.21	2.27	2.04	2.14	—
x_{col}^{top}	0.39	0.60	0.72	0.74	0.73	0.76	0.75	—

Compared with Nash equilibrium $(p^{\text{Nash}}, q^{\text{Nash}}) = (0.5, 0.1)$, as seen in (a) of Fig. 9, the probability p of the row agents in the simulation is considerably larger than the Nash prediction while the probability q of the column agents is slightly larger than it. As for payoffs, the payoffs in the simulation are Pareto superior to those in Nash equilibrium, and this feature is clearer for the top ten pairs of agents.

As seen in Fig. 8 and (b) of Fig. 9, although the row players' payoffs of human subjects in the experiments are considerably larger than the corresponding payoff in Nash equilibrium, the column players' payoffs in the experiments are slightly larger than the payoff in Nash equilibrium. Compare the result of the simulation with the experiments with human subjects. In the early generations of the simulation, it is observed that the payoffs of agents in the simulation are Pareto superior to those of the experiments. Finally the payoffs of the row agents are smaller than those of the experiments, and inversely the payoffs of the column agents are larger than those of the experiments. For the top ten pairs of agents, the payoffs of the row agents are almost the same as those of the experiments, and the payoffs of the column agents are obviously larger than those of the experiments. We will give the further consideration about comparison between behavior of agents in the simulation and that of human subjects in the experiments in the subsequent subsections.

4.4 Simulation Asymmetry

In this subsection, we investigate the relation between asymmetry of the payoff structures and the probability pair (p, q) of choices or the obtained payoff pair $(x_{\text{row}}, x_{\text{col}})$. In this simulation, agents are assumed to be risk neutral and to make a decision with no error, and then the parameters of risk attitude and error are set at $r = 0$ and $e_1 = 0$, respectively. Simulations with $a = 1, 1.5, 2, 3, 4, 5, 7, 9, 10, 20, 30$ are performed, and the results are shown in Figs. 10 and 11 where the probabilities of strategy choices and the payoffs received by agents are plotted for each value of a. Each point in the figures is an average of all the agents from generation 700 to generation 1000, i.e., the final 300 generations of the simulations, and these points are connected by a thick line. The averages of the top ten pairs are also provided, and they are given by a thin line. The probabilities of the Nash prediction and the corresponding payoffs are shown by dashed lines, and in Fig. 11, the payoffs of the random choices are also given. Moreover, the results of the experiments with human subjects are also depicted by gray rectangles. The result of Simulation Asymmetry for $a = 1, 2, 4, 7, 9, 20, 30$ is summarized in Table 4, and the trajectories of the payoff pair $(x_{\text{row}}, x_{\text{col}})$ with respect to changes of the parameter a are plotted in Fig. 12.

As seen in (a) of Fig. 10, the strategy choice probability p of the row agents increases with the parameter a, and in particular, when $a \leq 10$, the increase of the probability per unit of a is large. However, the growth of the probability slows down when $a > 10$, and it almost saturates when $a > 20$. Because the strategy choice probability of the row player in Nash equilibrium is always $p^{\text{Nash}} = 0.5$ for any a, the result of the simulation clearly differs from the prediction of Nash equilibrium. The difference between the strategy choice probability p of the row agents and the Nash prediction p^{Nash} also increases with the parameter a from $a = 1$ to $a = 20$, and when $a > 20$, the difference becomes almost the same.

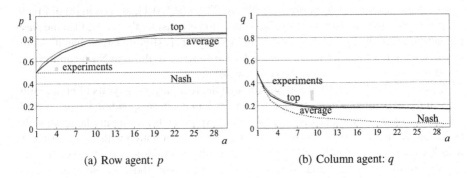

(a) Row agent: p (b) Column agent: q

Fig. 10. Probabilities p and q with respect to the parameter a

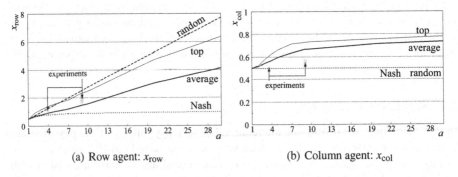

(a) Row agent: x_{row} (b) Column agent: x_{col}

Fig. 11. Payoffs x_{row} and x_{col} with respect to the parameter a

Conversely, the strategy choice probability q of the column player decreases with the parameter a as seen in (b) of Fig. 10. The decrease of the probability q per unit of a is relatively large when $a \leq 10$, and the probability q decreases little by little even if a becomes larger than 10. Because the strategy choice probability of the column player in Nash equilibrium is $q^{\text{Nash}} = 1/(a+1)$, the probability q in the simulation seems to follow the Nash prediction, but the probability q in the simulation is clearly larger than the Nash prediction q^{Nash}. The difference between them increases with the parameter a from $a = 1$ to $a = 20$, and when $a > 20$, the difference becomes almost the same.

Compare the result of the simulation with the experiments with human subjects. The strategy choice probability p of the row agents is larger than the experimental result p^{exp}, and the strategy choice probability q of the column agents is smaller than the experimental result q^{exp}, as seen in Fig. 10. Thus, the strategy choice probabilities of the simulation have a sharper feature than the results of the experiments, compared with the random choice.

For payoffs obtained by agents, because the payoff of the row player in Nash equilibrium is $a/(a+1)$, it converges at 1 as a grows. From (a) of Fig. 11, it is found that the payoff of the row agents increases linearly as a increases in the range of $1 \leq a \leq 30$. For the column player, the payoff in Nash equilibrium is always 0.5 regardless of the

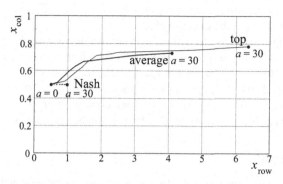

Fig. 12. Trajectories of the payoff pair in Simulation Asymmetry

Table 5. Result of Simulation Asymmetry

a	1	2	4	7	9	20	30
p	0.50	0.55	0.64	0.72	0.77	0.83	0.84
q	0.49	0.36	0.25	0.20	0.17	0.18	0.16
p^{top}	0.50	0.52	0.62	0.71	0.75	0.75	0.78
q^{top}	0.49	0.38	0.27	0.21	0.18	0.18	0.17
Nash p^{Nash}	0.50	0.50	0.50	0.50	0.50	0.50	0.50
Nash q^{Nash}	0.50	0.33	0.20	0.13	0.10	0.05	0.03
rand p^{rand}	0.50	0.50	0.50	0.50	0.50	0.50	0.50
rand q^{rand}	0.50	0.50	0.50	0.50	0.50	0.50	0.50
experiment p^{exp}			[0.534, 0.550]		[0.595, 0.643]		
experiment q^{exp}			[0.328, 0.336]		[0.241, 0.350]		
x_{row}	0.50	0.68	0.92	1.21	1.37	3.03	4.11
x_{col}	0.50	0.52	0.57	0.63	0.68	0.71	0.73
x_{row}^{top}	0.50	0.92	1.40	1.85	2.14	4.72	6.36
x_{col}^{top}	0.50	0.52	0.62	0.71	0.75	0.75	0.78
Nash x_{row}^{Nash}	0.50	0.67	0.80	0.88	0.90	0.95	0.97
Nash x_{col}^{Nash}	0.50	0.50	0.50	0.50	0.50	0.50	0.50
rand x_{row}^{rand}	0.50	0.75	1.25	2.00	2.50	5.25	7.75
rand x_{col}^{rand}	0.50	0.50	0.50	0.50	0.50	0.50	0.50
experiment x_{row}^{exp}			[1.013, 1.038]		[1.598, 2.256]		
experiment x_{col}^{exp}			[0.511, 0.5172]		[0.529, 0.574]		

magnitude of a. As seen in (b) of Fig. 11, the payoff of the column agents in the simulation is larger than that in Nash equilibrium. When $a \leq 10$, the increase of the payoff per unit of a is larger, and it increases little by little even if a becomes larger than 10. Compared with the experiments with human subjects, the payoffs of the row players in the experiments are between the average payoffs of the top ten pairs of agents and all the agents, which are indicated by 'top' and 'average' in the figures, respectively, and this means that the payoffs of human subjects in the experiments are in the payoff distribution of the simulation. For the column agents, the payoffs of the simulation are clearly larger than those of the experiments.

As shown in Fig. 12, the payoff pair in Nash equilibrium approaches $(x_{\text{row}}^{\text{Nash}}, x_{\text{col}}^{\text{Nash}}) = (1, 0.5)$ from $(0.5, 0.5)$. In the simulation, the payoff of the row agents starts from $x_{\text{row}} = 0.5$ in the game with $a = 1$ and it reaches $x_{\text{row}} = 4.1$ in the game with $a = 30$; it increases by 3.6. The payoff of the top ten row agents increases to $x_{\text{row}} = 6.4$ in the game with $a = 30$; it increases by 6.1. Thus, this results are clearly different from the payoffs in Nash equilibria, and this feature is clearly observed in the top ten pairs of agents.

4.5   Simulation Risk

In this subsection, we examine the relation between the risk attitude and the probability pair (p, q) of strategy choices or the payoff pair $(x_{\text{row}}, x_{\text{col}})$. We deal with a game with $a = 9$ where agents are assumed to make a decision with no error, i.e., $e_1 = 0$.

Simulations with $r = -0.5, 0.0, 0.5, 0.8$ are performed, and the results are shown in Fig. 13 where the trajectories of the payoff pair $(x_{\text{row}}, x_{\text{col}})$ are shown, together with the payoff pair in Nash equilibrium and the results of the experiments with human subjects. The payoff pairs close to the Pareto frontier are those of the top ten pairs of agents, and the trajectory of them moves from the lower right of the figure to the upper left. The payoff pairs in the middle of the feasible area are the average of all the agents, and the trajectory of them starts from $(2.5, 0.5)$ in the center of the figure and rotates to the upper left of the figure in a counterclockwise direction.

Fig. 14 shows the relation between the risk parameter and the strategy choice probabilities and the relation between the risk parameter and the payoffs of the agents. Each point in the figures is an average of all the agents in the final 300 generations of the simulation, and these points are connected by a thick line. The averages of the top ten pairs of agents are also provided, and they are given by a thin line. The strategy choice probabilities in Nash equilibrium and the corresponding payoffs are shown by dashed lines, and the results of the experiments with human subjects also are shown by gray rectangles. The corresponding numerical data is given in Table 6.

Because the utility functions of the row and the column agents are represented by $u_{\text{row}}(x; a, r) = x^{1-r} / a^{1-r}$ and $u_{\text{col}}(x; r) = x^{1-r}$, respectively, agents with utility functions with the parameter $r < 0$, $r = 0$ and $r > 0$ are risk prone, risk neutral and risk averse, respectively.

From Fig. 13 and Table 6, it is found that as the value of r increases, i.e., the risk attitude of agents varies from being risk prone to risk averse, the payoffs of the row agents increase and those of the column agents decrease, and this feature can be clearly observed in the result of the top ten pairs of agents as seen in (b) of Fig. 14. As for the strategy choice probabilities, as seen in (a) of Fig. 14, as the value of r increases, the strategy choice probability p of the row agents decreases and the strategy choice probability q of the column agents increases.

For a shift from being risk neutral $r = 0$ to risk prone $r = -0.5$, because as seen in Fig. 3 the marginal utility of the row agent around $x_{\text{row}} = 2$ decreases and that of the column agent around $x_{\text{col}} = 0.6$ slightly increases, it can be understood that the payoffs of the row agents decrease and those of the column agents show a slight increase.

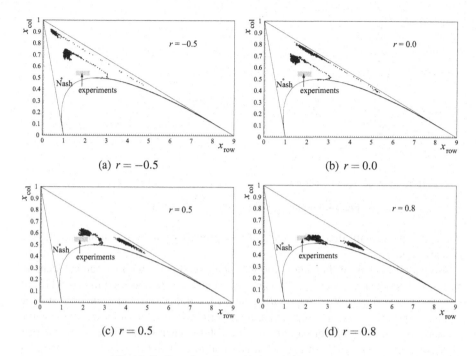

Fig. 13. Trajectories of payoff pairs in Simulation Risk

Table 6. Summary of Simulation Risk

r	−0.5	0.0	0.5	0.8	Nash	experiment
p	0.788	0.783	0.745	0.641	0.5	[0.595, 0.643]
q	0.132	0.171	0.277	0.371	0.1	[0.241, 0.350]
x_{row}	1.122	1.357	2.039	2.367	0.9	[1.598, 2.256]
x_{col}	0.711	0.686	0.610	0.536	0.5	[0.529, 0.574]
p^{top}	0.804	0.799	0.759	0.657	—	
q^{top}	0.123	0.177	0.302	0.395	—	
x_{row}^{top}	0.506	1.955	3.940	4.185	—	
x_{col}^{top}	0.896	0.756	0.520	0.482	—	

For a shift from being risk neutral $r = 0$ to risk averse $r = 0.5$, because of the opposite effect, it is observed that the payoffs of the row agents increase and those of the column agents slightly decrease.

We compare the result of the simulation and the Nash prediction. As seen in (a) of Fig. 14, both the strategy choice probabilities p and q of the row and the column agents in the simulation are larger than those in Nash equilibrium irrespective of their risk attitude, and the payoffs of the row and the column agents are also larger than those in Nash equilibrium. The difference between the payoff of the row agents in the simulation and that of the row player in Nash equilibrium is larger in the case of risk

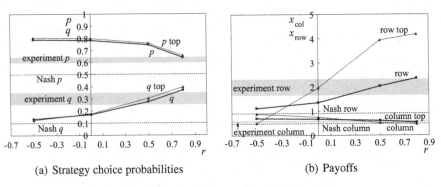

(a) Strategy choice probabilities   (b) Payoffs

Fig. 14. Summary of the result in Simulation Risk

averse, compared with that in the case of risk prone, and for the column agents, the reverse is true.

Compare the result of the simulation with the experiments with human subjects. In the cases of being risk prone $r = -0.5$ and risk neutral $r = 0.0$, as seen in (a) and (b) of Fig. 13, the payoff pair (x_{row}, x_{col}) starts from the random choice, after the row and the column agents obtain the payoff pairs Pareto superior to the result of the experiment, the row agents obtain the smaller payoffs and the column agents obtain the larger payoffs eventually. From (c) and (d) of Fig. 13, it is found that when the risk attitude of agents is risk averse, i.e., $r = 0.5, 0.8$, the payoff pair (x_{row}, x_{col}) also starts from the random choice and it converges in a set of points which are Pareto superior to the results of the experiments. The relation between the risk parameter and the payoffs of the agents converging in the long run is summarized in (b) of Fig. 14. As seen in the graph, the result of the case of being risk averse $r = 0.5, 0.8$ is similar to that of the experiments, compared with the case of being risk prone $r = -0.5$ or risk neutral $r = 0.0$. From this result of the simulation, we conjecture that the risk attitude of the human subjects in the experiments was risk averse.

4.6   Simulation Error

4.6.1   First Method: Adding the Error Term

In the first method for implementing the error in decision making, the error is characterized by the magnitude e_1 and the direction α of the error, and it is represented by (9) and (10). In this method, the degree of the error increases with the value of e_1. We deal with a game with $a = 9$ where agents are assumed to be risk neutral, and then the parameter of risk attitude is set at $r = 0.0$. Simulations with $e_1 = 0.1, 0.2, 0.3$ are performed, and the results are shown in Fig. 15 including the result of the simulation with no error. These graphs are depicted in the same way as the graphs for Simulation Risk in the previous subsection.

Fig. 16 shows the relation between the error parameter and the strategy choice probabilities and the relation between the error parameter and the payoffs of the agents. Graphs in Fig. 16 are also depicted in the same way as those for Simulation Risk in the previous subsection. The corresponding numerical data is given in Table 7.

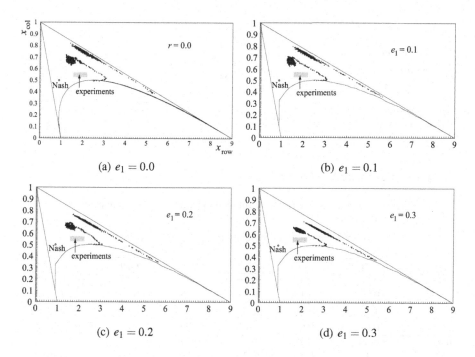

Fig. 15. Trajectories of payoff pairs in Simulation Error with the parameter e_1

From Fig. 15 and Table 7, it is observed that as the value of e_1 increases, i.e., the error of agents in decision making becomes larger, the payoffs of the row agents increase and those of the column agents decrease, and this feature can be clearly observed in the result of the top ten pairs of agents as seen in (b) of Fig. 16. As for the strategy choice probabilities, as seen in (a) of Fig. 16, as the value of e_1 increases, the strategy choice probability p of the row agents decreases and the strategy choice probability q of the column agents increases. However, when e_1 is smaller than 0.2, this feature is not clear, and the difference between the results of $e_1 = 0.2$ and $e_1 = 0.3$ is larger than the difference between the results of $e_1 = 0.0$ and $e_1 = 0.1$. This fact can be account for by the conjecture that because the learning mechanism of the agent-based simulation model works well, small errors have an insignificant effect on the performance of the agents.

As seen in Fig. 16, both the strategy choice probabilities p and q of the row and the column agents in the simulation are larger than those in Nash equilibrium irrespective of the magnitude of the error parameter e_1, and similarly, the payoffs of the row and the column agents are also larger than those in Nash equilibrium.

Compare the result of the simulation with the experimental data. As seen in Fig. 15, the payoff pair $(x_{\text{row}}, x_{\text{col}})$ starts from the random choice, and after the row and the column agents obtain the payoff pairs Pareto superior to the results of the experiments, the row agents obtain the smaller payoffs and the column agents obtain the larger payoffs eventually. The final situation of the simulation is shown in (b) of Fig. 16, and as seen

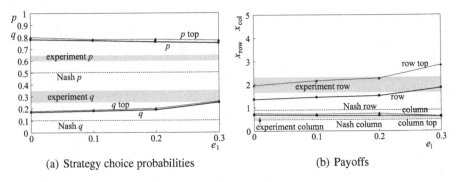

(a) Strategy choice probabilities   (b) Payoffs

Fig. 16. Summary of the result in Simulation Error with the parameter e_1

Table 7. Summary of Simulation Error: the first method

e_1	no error	0.1	0.2	0.3	Nash	experiment
p	0.78	0.77	0.76	0.75	0.5	$[0.595, 0.643]$
q	0.17	0.18	0.19	0.25	0.1	$[0.241, 0.350]$
x_{row}	1.36	1.45	1.52	1.85	0.9	$[1.598, 2.256]$
x_{col}	0.69	0.67	0.66	0.63	0.5	$[0.529, 0.574]$
p^{top}	0.80	0.78	0.78	0.77	—	
q^{top}	0.18	0.19	0.20	0.26	—	
x_{row}^{top}	1.96	2.16	2.27	2.84	—	
x_{col}^{top}	0.76	0.72	0.76	0.63	—	

in this graph, the result of the case of $e_1 = 0.3$ is more similar to that of the experiments, compared with the other cases. This fact suggests that the error in decision making is a possible cause explaining the behavior of human subjects in the experiments.

4.6.2   Second Method: Modifying the Output

In the second method for implementing the error in decision making, the output of the neural network is modified so as to approach the random choice, and this method is characterized by a magnitude e_2 of the error. The error represented by (11) and (12) increases with the value of e_2, and as seen in Fig. 4, the probability p or q comes closer to 0.5 as the value of e_2 grows. We deal with a game with $a = 9$ where agents are assumed to be risk neutral in a way similar to the simulation with the first method of representation of the error. Simulations with $e_2 = -0.5, -0.2, -0.1$ are performed, and the results are shown in Fig. 17 including the result of the simulation with no error.

Fig. 18 shows the relation between the error parameter and the strategy choice probabilities and the relation between the error parameter and the payoffs of the agents, and the corresponding numerical data is given in Table 8.

As seen in Fig. 18 and Table 8, the result of the second method is similar to the result of Simulation Error with the first method, and the observed feature can be seen more

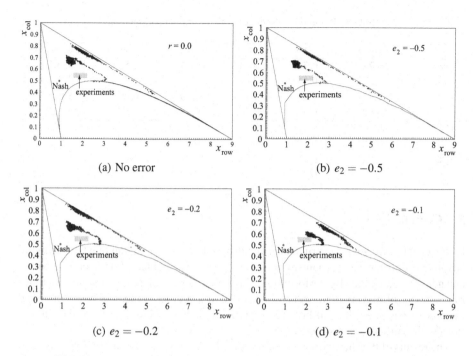

Fig. 17. Trajectories of payoff pairs in Simulation Error with the parameter e_2

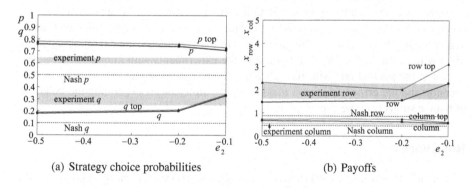

(a) Strategy choice probabilities   (b) Payoffs

Fig. 18. Summary of the result in Simulation Error with the parameter e_2

clearly in this method, compared with in the first method. Furthermore, it is observed that the result of this method is more similar to that of the experiment, and therefore from this fact, it is thought that as the representation of the error in this game the second method may be more suitable than the first one.

Table 8. Summary of Simulation Error: the second method

e_2	no error	−0.5	−0.2	−0.1	Nash	experiment
p	0.78	0.76	0.74	0.71	0.5	[0.595, 0.643]
q	0.17	0.18	0.20	0.33	0.1	[0.241, 0.350]
x_{row}	1.36	1.46	1.56	2.29	0.9	[1.598, 2.256]
x_{col}	0.69	0.67	0.64	0.58	0.5	[0.529, 0.574]
p^{top}	0.80	0.78	0.76	0.73	—	
q^{top}	0.18	0.19	0.21	0.34	—	
x_{row}^{top}	1.96	2.32	2.02	3.12	—	
x_{col}^{top}	0.76	0.73	0.76	0.62	—	

5   Conclusions

In this chapter, we have constructed an agent-based simulation model in which artificial adaptive agents have a mechanism of decisions and learning based on neural networks and genetic algorithms. By performing several simulations, we have examined the long-run behavior of players in the generalized matching pennies game, and have compared the predictions of Nash equilibria, the experimental data, and the results of the simulations with artificial adaptive agents.

The results of the simulations are summarized as follows. In the simulations based on the adaptive behavioral model, the agents do not behave like the Nash prediction, and in this sense the results of the simulations support those of the experiments with human subjects. From the results of the simulations, we conjecture that in games with experienced players the column player would obtain more payoff at the sacrifice of a small loss of the row player, and with augmentation of the asymmetry, the payoff obtained by the row player increases. With reference to the experiments with human subjects, the results of the simulations with risk attitude and error suggest that the property of risk aversion and the error in decision making are possible causes explaining the behavior of human subjects in the experiments.

References

1. Axelrod, R.: The evolution of strategies in the iterated prisoner's dilemma. In: Davis, L. (ed.) Genetic Algorithms and Simulated Annealing, pp. 32–41. Pitman, London (1987)
2. Axelrod, R.: Advancing the art of simulation in the social sciences. In: Conte, R., Hegselmann, R., Terna, P. (eds.) Simulating Social Phenomena, pp. 21–40. Springer, Berlin (1997)
3. Andreoni, J., Miller, J.H.: Auctions with artificial adaptive agents. Games Econ. Behav. 10, 39–64 (1995)
4. Banerje, B., Sen, S.: Selecting parters. In: Parsons, S., Gmytrasiewicz, P., Wooldridge, M. (eds.) Game Theory and Decision Theory in Agent-Based Systems, pp. 27–42. Kluwer Academic Publishers, Boston (2002)
5. Chellapilla, K., Fogel, D.B.: Evolving neural networks to play checkers without expert knowledge. IEEE Trans. Neural Netw. 10, 1382–1391 (1999)
6. Conte, R., Hegselmann, R., Terna, P. (eds.): Simulating Social Phenomena. Springer, Berlin (1997)

7. Dorsey, R.E., Johnson, J.D., Van Boening, M.V.: The use of artificial neural networks for estimation of decision surfaces in first price sealed bid auctions. In: Cooper, W.W., Whinston, A.B. (eds.) New Directions in Computational Economics, pp. 19–40. Kluwer (1994)
8. Downing, T.E., Moss, S., Pahl-Wostl, C.: Understanding climate policy using participatory agent-based social simulation. In: Moss, S., Davidson, P. (eds.) Multi-Agent Based Simulation, pp. 198–213. Springer, Berlin (2001)
9. Epstein, J.M., Axtell, R.: Glowing Artificial Societies. Brookings Institution Press, Washington, D.C. (1996)
10. Erev, I., Rapoport, A.: Coordination, "magic," and reinforcement learning in a market entry game. Games Econ. Behav. 23, 146–175 (1998)
11. Erev, I., Roth, A.E.: Predicting how people play games: reinforcement learning in experimental games with unique, mixed strategy equilibria. Am. Econ. Rev. 88, 848–881 (1998)
12. Duffy, J., Feltovich, N.: Does observation of others affect learning in strategic environments? An experimental study. Int. J. Game Theory 28, 131–152 (1999)
13. Goeree, J.K., Holt, C.A.: Ten little treasures of game theory and ten intuitive contradictions. Am. Econ. Rev. 91, 1402–1422 (2001)
14. Goeree, J.K., Holt, C.A., Palfrey, T.R.: Risk averse behavior in generalized matching pennies games. Games Econ. Behav. 45, 97–113 (2003)
15. Holland, J.H., Miller, J.H.: Adaptive intelligent agents in economic theory. Am. Econ. Rev. 81, 365–370 (1991)
16. Kahneman, D., Tversky, A.: Prospect theory; an analysis of decision under risk. Econometrica 47, 263–291 (1979)
17. Karmarkar, U.S.: Subjectively weighted utility; a descriptive extension of the expected utility model. Organ. Behav. Hum. Perform. 21, 61–72 (1978)
18. Leshno, M., Moller, D., Ein-Dor, P.: Neural nets in a group decision process. Int. J. Game Theory 31, 447–467 (2002)
19. McKelvey, R.D., Palfrey, T.R.: Quantal response equilibria for normal form games. Games Econ. Behav. 10, 6–38 (1995)
20. McKelvey, R.D., Palfrey, T.R., Weber, R.A.: The effects of payoff magnitude and heterogeneity on behavior in 2×2 games with unique mixed strategy equilibria. J. Econ. Behav. Organ. 42, 523–548 (2000)
21. Moss, S., Davidson, P. (eds.): Multi-Agent Based Simulation. Springer, Berlin (2001)
22. Nishizaki, I.: A general framework of agent-based simulation for analyzing behavior of players in games. J. Telecomm. Inf. Technol. 4, 28–35 (2007)
23. Nishizaki, I., Katagiri, H., Oyama, T.: Simulation analysis using multi-agent systems for social norms. Comput. Econ. 34, 37–65 (2009)
24. Nishizaki, I., Ueda, Y., Sasaki, T.: Lotteries as a means of financing for preservation of the global commons and agent-based simulation analysis. Appl. Artif. Intell. 19, 721–741 (2005)
25. Niv, Y., Joel, D., Meilijson, I., Ruppin, E.: Evolution of reinforcement learning in foraging bees: a simple explanation for risk averse behavior. Neurocomputing 44-46, 951–956 (2002)
26. Ochs, J.: Games with unique, mixed strategy equilibria: an experimental study. Games Econ. Behav. 10, 202–217 (1995)
27. Parsons, S., Gmytrasiewicz, P., Wooldridge, M. (eds.): Game Theory and Decision Theory in Agent-Based Systems. Kluwer (2002)
28. Rapoport, A., Seale, D.A., Winter, E.: Coordination and learning behavior in large groups with asymmetric players. Games Econ. Behav. 39, 111–136 (2002)
29. Roth, A.E., Erev, I.: Learning in extensive form games: experimental data and simple dynamic models in the intermediate term. Games Econ. Behav. 8, 163–212 (1995)
30. Sichman, J.S., Bousquet, F., Davidsson, P. (eds.): Multi-Agent-Based Simulation II. Springer (2003)
31. Sundali, J.A., Rapoport, A., Seale, D.A.: Coordination in market entry games with symmetric players. Organ. Behav. Hum. Decis. Process. 64, 203–218 (1995)

Author Index